Life Through A Device

-We Won't Be Distracted By Comparison If We Are Captivated With Purpose.

-Bob Goff

Life

Through

A

Device

Unlock your Mind Rather than your Phone

William Civitillo

Life Through A Device

I want to dedicate this book, of course, to my family and friends. If it weren't for you guys always on your phones, the idea would have never come to mind.

To my Grandmother up in Heaven, you had taught me so much when you were here. I think about your lessons and guidance day in and day out.

To Jeff Sarri, I am forever thankful to be a student of yours. You have taught me the most important lessons in life. I am forever grateful.

One last dedication goes to my childhood. Filled with imagination and wonders from playing outside instead of apps, and how many likes I can get on a picture.

Preface

Why do people today worry more about how many likes they get on their profile pictures than what they are going to eat for dinner? The answer is simple. The human mind is now trying to adapt to the ever-evolving age of technology. The world around us is changing at a dramatic rate. We dress differently, talk differently, listen differently, and most importantly - we act differently. Chunks of data are replacing our knowledge, and our intuition is now based-off of deadlines. The simple things in life seem to be slipping away from us. I feel as though we aren't yet computed to go hand-in-hand with the technology bestowed upon us. People all around the world eat, sleep, and breathe social media. Adults and children are quickly falling for the trap hidden in social media; driven by their phones, this trap being "Drowning in a sea-of-self."

Even with all the terrifying diseases that come from inadequate nutrition and lack of physical exercise, Americans seem to be adapting to more of a sedentary lifestyle. Physical health is not only going south, but also social cues and mental defects may arise. Society, as we know it can be looked up and searched on the phone. However, did you know if you don't pick up your head and look around, you can miss the most important and most special moments to ever occur in your life?

I'll ask you a question right off the bat. Keep you on your toes with enraging anticipation on what I am about to give you. How would you define "Social Media," in one sentence? You might have thought to connect with friends or kill some time. I define "Social Media" as a simulated and inviting space to impress or obtain attention from those online. This may be hard to understand at first, hell, I barely understand it. But this book may show you how and why a phone enables a person to be someone who they are not. Living your life through a device and living off of others compliments and pity is not a way of living life to the fullest. No one cares if you are on vacation or got a new puppy. "Attention" is the solid base of social media.

If you are constantly looking at your phone, researching ridiculous crap, and laughing at "memes," then this may be a short novel for you. However, if you don't think you have the mental capacity, then go back to social media. Better hurry, your "friends" are waiting for you. Aye, see, I knew you would stay. Well, we have lots to talk about, so, put your phone to the side as we delve into the novel. Sit back, relax, and unchain that mind of yours. Be ready to enter the world of useful book knowledge. Don't get too comfortable, though! Remember, you need attention span when you read.

Synchronicity!

An Important Lesson; From a Simple, But Brilliant Professor

My favorite professor from my undergraduate college program taught me a valuable lesson that I will never forget -A lesson that I will hold deeply and apply daily. I remember the first day walking into his class. This guy was sitting on his desk, barefoot. I was thinking to myself, "Oh, here we go. I got myself another wack-a-do professor." However, that changed ever so quickly when he asked the class, "So, who is ready to teach?" We all sat in our seats, stunned, and let our emotions flourish within us. Not one student said a word. Then, out of nowhere, Mr. Jeff Sarri blurts out the word, "Synchronicity!" The class was still speechless, trying to understand if this guy is a professor or someone off the street. From this moment, my life within the classroom and out was changed drastically. I also should note that this class was called Behavior Change. Mr. Saari expressed the word, 'Synchronicity,' like the simple things we all need in life. In other words, those "how did that just happen!" moments; the moments you will not forget, and the moments you will cherish forever. The moments that bring out the good in you and tell you that life has its ups. He taught me to embrace life and take in the simple things, which are the most soothing. The homework assignments consisted of finding these frequent moments. But here was the trick, you had to remember these moments in the long run.

After initially putting some thought into it, I realized I didn't feel these "coincidences" happen very often. However, going through Jeff's class, I have certainly learned the "hows," "whats," and "wheres" that define the science of synchronicity. These days, there are distractions left and right that make you miss your moments. Staring at your phone will never give you that feeling of joy. By continually looking down, you'll miss out on those synchronistic moments, and more importantly, life. For example, say you have a song playing in your head. An hour later, that same song is playing on the radio on your way to work–boom-synchronicity. It may not be a life-changing moment, but it makes you ponder on how the hell that happened. Don't confuse this with Deja Vu because if you experience this many times, you may be schizophrenic instead of synchronistic and should probably get checked out. No, I'm only kidding, but if you stare at your phone for days on end, you may develop it.

Mr. Sarri always had a good spirit and a huge smile on his face, whether it be lecturing or was outside of the classroom. He also always had the same morning ritual. He would always go to the downtown coffee shop to purchase green tea with lemon. Then, of course, he would try to spur up a conversation with a person in there, and would never make the conversation about himself. I always envied his positive attitude towards the world, and I still do to this day. One day out on campus, I ran into him. Something came over my mind to ask him this edging question, "Hey Jeff, how can you always be in such an uplifting mood?" He laughed and said, "I seem to get that question, at least, once a week. This is why I am a teacher and a motivational coach. It is my life passion to help people get through the roughs and downs." He continued into more detail and told me a life-changing story that he doesn't tell many. He says, "Back when I was about your age, (which was 21), I tried to hang myself. Luckily I failed and ended up in the hospital. From that moment, when I opened my eyes once again, realizing where I was, I knew I had a life-calling." I was stunned when he told me this. Who would've thought that such a happy go get'em guy tried to commit suicide in the past? He then went on to say, "This is why, and many of the reasons why you can't take life for granted. Not even for a second. Take in every moment and embrace it, whether it is positive or negative." "I have learned a great deal through my struggles, and I learned to harness those difficulties and shed some light on those who feel powerless or unstrained like I once was."

Jeff's life coaching skills and extensive knowledge of the human mind has undoubtedly rubbed off on me quite a bit. He has helped many in the past, whether it be students or adults going through some crisis. He says expressing your values and beliefs is the most important. "This is why expression is hard to get out of anybody these days." I stared into his ocean blue eyes like looking into a deep dark abyss, patiently waiting for what he was going to say next. "Phones and all of these mind-numbing devices are keeping people hindered within a shell, that not even I can break into, and I have been doing this for years." He continued, "Phones and the social media craze that comes with them are like a gateway for people to find peace within themselves when in reality, they have a weak mind and obstructed expressions and views about themselves and others." I replied wittingly, "Well, now that you mention it, I do see many students on campus walking to class usually looking directly down into their phone. I'm always wondering what they are looking at, or what they are waiting for." Jeff said, with a big pearly white smile on his face, "Exactly, you got it right on the money, Will." The short, but the eye-opening conversation ended, and he went on to his way to his next class. He waved in the distance with his back turned to me, saying, "See ya around Will, study a tree instead of your phone, he shouted." I laughed and remarked back, "Right back at ya, Jeff!"

As we walked off, I still went about my day. I walked back to my place a few blocks away from campus. I lived with three other students whom I am still very close to. As I opened up the front door, I walked into a very dim living room with all the blinds closed. It reminded me of bat cave since it was only about two in the afternoon on a sunny Fall day. One of my roommates was sitting on the couch. Take a wild guess as to what he was doing. No, not reading a book, but blankly looking into a bright phone screen, waiting for the next big thing to happen. I even noticed that as I walked in, he didn't even bother to pick up his head to see who it was. I said, "Hello? Anybody in there?" He finally looked up, and said, "Oh hey, what up dude?" Then dropped back into his phone. I could have straight-up robbed the house if I didn't live there. As I walked up to the stairs into my room, it felt like Jeff was talking to me, "Now, you know what I mean, Will?" It all started to connect even though it still felt confusing as to why Jeff couldn't break into someone trapped behind a phone.

The next day, I came home to the same scenario. My roommate was in the same spot on the couch with his phone in his hands. I asked him, "Is this what you do all day, every day?" He looked at me with confusion and said: "What do you mean?" I replied hastefully, "I mean that damn phone of yours! Do you ever put it down?" He was shocked at the way I reacted, and he quickly put his phone down, he said, "Jeez relax, what's up with you?" I said, "I think you need help bud, that phone of yours isn't a habit," with an alerted look on my face, "Its an addiction." He then gave me a dirty look, flipped me off, and said, "Go screw yourself." I'm guessing you could already tell what his next move was.

Jeff's lesson was hitting me in the gut now. As I looked around the campus at this time of the year, everything was gleaming with color, from the beautiful flowers to the ever-changing pigments of leaves on trees. New Hampshire in autumn is a sight to see. I'm starting to visualize the world as a whole instead of looking at it through screens on devices. Picking up your head makes you realize who you are, and who you want to become in life.

May sound like an idiotic truth, but remember, it's the simple things that keep us going. Observing other students on campus, I noticed that some are walking into each other and tripping on their own two feet because they can not look ahead of themselves. Not taking in the beauty and awe of what our campus offers to bring. Hey, I am not even going to lie to you guys, it is hard to stop looking at your phone. It seems like it is the social norm in today's society. The generation that I am a part of is booming with the new technology age. All of us are still relatively young and don't know our exact path in life yet. Lessons in life will come and go, but those lessons you remember from the beginning of your life will stay at the base of your beliefs and values throughout the rest of your life. Lessons that not only will change you from the inside out but will change your perception of life as a whole. Let me tell you this, those lessons are hard to find, and you will not find those lessons within a phone. I was blessed to have a professor, mentor, and a friend like Mr. Sarri. A professor who didn't give tests or exams, but instead, gave trials through the mind.

As a senior in college, I was packed with knowledge from teachers from the 1st grade, and it felt easier to discipline myself and learn from mistakes. Giving up bad habits that would potentially jeopardize my upcoming future felt easier. Every day, I feel like I am on a mission to better myself and others. Even though I am still young, the one true code I praise and hold onto deeply is to give to others before you receive. This can be anything: money, various items, or even better, knowledge. Give back to the community what it has given to you, and the luxury of happiness will be right around the corner. Just do to others what you would want them to do to you.

At the start of my junior year, I decided to give my time and volunteer at the local community kitchen every Sunday morning. I mean hell, I was about hung-over every time I showed up at 9:30 am but let me tell you this. It felt a lot better lending a hand then getting another hour of sleep. I won't even lie; the reason why I started volunteering at the community kitchen was to spice up my resume. However, that inadequate mindset changed once senior year came, and Jeff was my new mentor. Volunteering at the kitchen was on a whole new spectrum. I did not only gave my time and effort, but now, I was spending my time getting to know the people that ate there every Sunday morning.

A lot of their stories were heartbreaking, and also heartwarming. I will never forget this one fellow. His name was Gene. He was a bit on the older side and had his life strapped to his bike. He would come in every Sunday when the breakfast line was about to close. Every time, he would have sealed containers to take extra food with him. He would eat his first meal of the day and pack others up. This was the rest of the food supply for his day. He made me think how good I had it growing up and at school. I never felt more blessed to have a life with a roof over my head and food constantly on my plate. Working at the kitchen made me realize to never take anything for granted. If I'm having a bad day, or even week, I stop and think about what I have that others more unfortunate than me don't.

One day, I decided to break out of my comfort zone and sit and talk to Gene. Of course, I didn't know his name at the time, but the first conversation I ever had with him felt more reasonable than a conversation I had with my ex-girlfriend. Small talk turned into life stories in a matter of minutes. Gene was from Idaho, and his parents passed when he was young. He has two sisters and a brother who he still keeps in contact with. They all live in New England. He told me that he had had bad luck all of his life. His ex-wife completely abandoned him, and he struggled financially. He says that his depression was so strong that he thought about ending it several times, but he always kept on his toes. I not only saw him in the kitchen but also around town and in the campus library where he would always be reading a different book. Many others were eating in the kitchen stuck in the same position as Gene, but all had different stories.

One of the most inspirational people I met in the kitchen was a woman named Christina. She had always been down on her luck, ranging from constant job changes to losing her arm in a motorcycle accident. She was a very kind-hearted woman who always had a smile on her face. She always asked how everyone, including myself, was doing. She was always thankful and very generous. As time went on, she eventually got a prosthetic arm. I couldn't believe it. What was once a nub was now a full and mobile arm. She got a job at the local grocery market and was able to rent out a place of her own. A truly inspirational story that I still think about.

I have countless uplifting stories about those people that many students from my school would never approach, never mind converse with. However, these are the people from whom one can learn the most from because of their different perspectives. Opening your mind and heart is the way humans learn and excel. Limiting yourself to those only in a specific social group can only lead you so far. Expand your horizons, and you will thank me later.

You may be thinking to yourself right now, "Did Mr. Sarri give Will an essay to write for a grade?" Or maybe you aren't thinking of that at all. What I am saying is that this is no boring essay about today's world. Trust me, and I would not torture you with some crap like that. This novel is here for some laughs, knowledge, and for you to limit your phone usage to attain new priorities in life. But, let me tell you something, the truth hurts. You can either accept it or deny it. I couldn't care less what you do with the truth. I am still going to write this book for you guys. You can keep reading or look at what is new on social media. Don't worry, I understand. If you have read this far without stopping, it has been at least a solid 15 minutes. There is bound to be something new on there. Maybe a new meme! Or a new celebrity relationship! And hey, off the record, you finished the first chapter, pat yourself on the back. People these days may go out and buy a book, but reading it may be a little too far fetched. Alright, let us get back on track and dissect that lesson that Mr. Sarri instilled in me throughout the fall semester. I still ponder on the lesson today to sustain and improve its eye-opening gloriousness.

Infusing That Lesson

Jeff taught me this vital lesson at a breaking point in my life. The point where I was about to graduate from college and turn the page to the next chapter. It was very bittersweet graduating from a college that I held onto so dearly, but it was a happy and well-needed ending. My binge drinking days and constant fighting with roommates is now over. It was time to start my adult life. Little did I know, the warning signs were bestowed on me months and even years before graduation that the adult life has no set schedule or direction. I was, of course, agreeing with the valid statement, but in my head, I was saying, "I can not wait never to pick up a calculator, or solve a scientific problem ever again." I regret not taking those words seriously because I was expecting a path after college. However, I now realize that you have to make your path. The medication of college was starting to wear off, and now, the pain and the constant worry of the real-life were beginning to settle in and stay permanently. It was an overwhelming sadness to come to this realization.

You're probably asking yourself, " Why is this guy bitching about his life after college? We all struggle from it." Well, here is the lesson - the simple and true lesson of synchronicity. If you felt the same struggles or you think you soon will, you just felt the power of synchronicity. The verse of synchronicity should be replayed as much as possible for ongoing subtle emotional bursts. Synchronicity builds your mind not only to observe your surroundings but also absorb them as well. Do you know how the world is a mean and vicious place? Well, now you have the secret, mental weapon that is infinitely more powerful than any other influence out there.

To this day, this simple, yet, the outlandish lesson still seems to slip from me. However, every single day, I think of what Coach Jeff has bestowed upon me. (From now on, whenever I bring up Mr. Sarri, I am going to refer to him as Coach Jeff.) Coach Jeff never taught me how to find the sum of x, or solve the momentum of gravity, but he taught me to take life and not take it for granted. I'm sure everyone has heard that before, but the way Jeff has taught this lesson to me, time and time again, I will never forget its value. I am forever thankful for having such an intelligent life professor during my senior year of college. If I had taken his class sooner, his knowledge would have probably slipped by now for the simple fact that I was more naïve when I was younger. I feel like it is my duty now to pass his knowledge on to all you readers.

Coach Jeff has helped many in the past, even my mess with my ex-girlfriend. I was dating her around the same time I had Jeff as my professor, in my fall semester. Now that I think about it, if it weren't for Jeff, I would have probably jumped off a bridge at that time. I enjoyed the relationship in the beginning. I mean, who doesn't? But as time goes on, things changed, mindsets run rapid, and emotions get the best of us. I soon realized that I was hiding a depressive side of me when I was in that relationship, and when that was finally released, things started to head south. It wasn't her fault or mine. Eh, maybe it was her fault a bit, but hey, we are all human.

Well, anyway, when we broke up, she was going through a lot of family issues that I will, of course, keep private. I mean hey, maybe if you stick around, I will release a couple of beasts. No, I am just kidding, I would never do such a thing to such a caring woman. Getting back on track, I asked Jeff to see if he wanted to help. At this point, not even her closest friends could help her; she needed professional help. So, I stepped in, even though we mutually hated each other at the time, I chose to become the bigger person and intervened. Naturally, Jeff, the most heartful guy I know agreed to give her a couple of sessions. She was excited to spew all of her built-up emotions to Jeff. Even though she had no idea who he was, she said the conversation went very well, and it helped a lot. She also said, "It's like he became my best friend in a matter of five minutes." I nodded my head and said, "See, it all works out when Jeff is involved." She then said, "Jeff brought you up a couple of times as well. He even mentioned that you were a very talented writer." I was baffled at the statement, and I jokingly said, "Maybe I should write a best selling book about what a bitch of a girlfriend you were," laughing out loud in her face. She walked off while flipping me off. And you know what, I completely respected it. Also, I respected what Jeff said about my writing. It means a lot when someone you look up to appreciates what you do.

You are probably thinking to yourself, "What the hell does this have to do with phones?" I'll tell you the truth. This little segment has nothing to do with phones or even my ex-girlfriend. She was just an example of the bigger picture. The picture of life, and how it can be ever so dandy. Life entails relationships, break ups, achievements, and downfalls. The wicked part about life is that it never stops. It is always taking sharp curves and U-turns. You may think life stops when you set your mind on something - say maybe your phone - the world stops revolving? No, of course not. This constant distraction shouldn't and should never be intervening with your life. Memories don't come from phones and its attachments. It comes from living. It is living the life you were born to live. Take on your fears and worries with minimal distraction. Use your phone to a negligible amount, and you will see what I mean. The powers of synchronicity and those from within you will be unlocked, and you could use them to an advantage over many that are weak-minded.

I want you to ask yourself this. Do you believe that capturing pictures is more important than capturing moments? You may answer this by saying to yourself, "But capturing the photo will always be an object in a frame that I will always have." You are missing the point. Capturing moments is the actual thing you put into the frame in your mind. Taking a picture on your phone can be nifty at the moment, but in a week from now, you will completely forget to even look at that picture or think about it. Every moment you experience can speak infinite more words than any picture can.

For all this to work, you have to follow your theory. By that, I don't want you to go off of my beliefs but develop your own. You don't have to agree with everything I have to say, but certainly, take in the knowledge I obtained from my guides in life. Whatever you want in life isn't just going to show up at your doorstep tomorrow, or even next year. Developing a plan is the hardest part, and I am going to show you how. However, don't think I am going to shoot a random fabulous idea into your head. That is your job and your job only. I'm here to inspire through my writing to show you how you can expand on that idea, and never give up.

Another phrase we all use too frequently is, "given up," and there is usually an "I have" before those words. People, including you and me, use it way too much, even for the simplest of tasks. Just imagine if every person completed his or her task every single day. Even if there are multiples mistakes, at least, he or she didn't give up. People give up for some of the ridiculous tasks, such as doing laundry or doing math homework. There are even people out there giving up on sex, not because they can't get any of it, but because they can't finish. Oh no, c'mon Will, those people are just lazy. You know what? I don't believe that laziness or giving up correlate. Those that are lazy are the ones who never started. The ones who give up have started something, but stop the process because it either got too hard or too time-consuming. So now, I am going to lay onto you another lesson that Jeff has taught me - the Transtheoretical Model. I'm sure most of you have heard of it before and some of you probably already know the steps. However, I bet you didn't have Jeff to extend your knowledge in this model.

First, I am going to lay out the steps so you can fully grasp the concept. The initial step is pre-contemplation, which essentially is someone not willing to change. I say these people are entirely unaware of their life and their existence as a whole. You don't want to be this guy because he is not going anywhere fast, even less than the speed of a snail with 50-pound weights underneath its shell. Maybe these people are already happy and in a state of bliss, until a bomb known as reality strikes with no mercy. The future will come with frightening surprises and catch these people off guard. They take for granted the pros in their life and put no emphasis on the cons. Depression will set in, and it will be an immense feeling until that person decides it is time for a change.

The second step is when the person knows and plans on a life change. This step is called contemplation. Just to let you guys know, about 80% of the population is stuck within these first two stages. Those in the second stage are a hundred steps ahead of the pre-contemplators. People in the second stage currently notice problematic factors in their lives. They intend to start improving this behavior, but can't seem to find the right spot to pounce upon. They recognize both the pros and cons in their life, but still have that feeling of ambivalence lingering around. These feelings contradict their ideas, and improvement is hard to find. That feeling of doubt is the main reason why people give up. There are many others, such as quitting too soon, or not having the patience, but that feeling of doubt takes over the mind and will suppress self-promotion. I do feel apathy for those in this stage because all they need is that one push. That one push is just as hard, if not harder than putting the first foot forward.

Once somebody is in the right direction and ready for a change, we can go on to the third stage of the model. Once at this step, preparation, the person realizes the flaws and is ready to improve and act upon them. They're ready to make the change and now need the tools to help better themselves. These next small steps will help someone head towards completing the model and themselves as well. I also like to call this stage "determination phase." This phase will lead people to an overall healthier life. This person is ready and prepared to take on the beasts ahead whatever they may be.

Fast forward one month, and this person is in the action phase of the model. The scary part of this phase is that the person may give up more quickly because they don't see any change in themselves, or results don't last. The intention to keep moving forward can either grow or deteriorate in a person. If that person does give up, the model may start again from the 2nd step, where the person knows a change has to be done but doesn't know how to approach it. On the bright side, those who do stick with the plan and keep pushing will soon find themselves six months later in the next phase, where they don't even have to exhibit the problems anymore. Instead, they will modify their problematic behaviors and turn them into knowledge gaining and positive ones.

The fifth phase is known as the maintenance phase. In this stage, the person fully understands their behavior and has sustained benefits from changing that behavior. And guess what? These people will maintain these behavior changes and take note when something is not going according to plan. Instead of whining about their problems, they will act to resolve issues along the road of acquiring the new healthy behaviors they have been working so hard to achieve. There will be many mistakes made throughout the five steps I have spoken about, but the margin of error will never outweigh reaching the goals you set out. Throughout this prolonged phase, people will work to prevent a relapse of their old habits. After time and many months of stress, the sixth phase will be right around the corner.

The sixth and final step of the model is known as the termination phase. In this phase, the person has been successful in the maintenance phase for about five years. That may seem like a lot of time, but once you know relapse will never happen, you have completed the model, and your healthy behavior will stick with you for life. Believe it or not, not many people make it to this stage. It is easy to relapse back into old habits, and pain will come of this. But the reward and knowledge obtained throughout this journey will change you forever, and the person you once knew before is a distant memory. The desire to go to old habits and unhealthy behaviors has been eliminated.

I have to say; I applaud those who make it to the maintenance phase. It is also crazy to think that 80% of people are stuck still in those first two phases. This model is here as a guide to help unlock yourself and benefit you and those around you. You have the power to change. If you don't want to, I respect it because you are your person. Whatever you want to do in life is up to you and only you. However, I must add, phones that are always in our pockets change the game of following the model accordingly.

I remember Jeff using an example of his experience going through the phases. He started to get into the gym to better himself and his everyday life. He loves the stress relief of "Pumping irons," he always said. I couldn't agree more. I feel like the gym is the only place where I could get lost. Just run wild and free amongst the dumbbells. Anyways, Jeff's style of training is more cross fit based. I don't know much about his style, but whenever I saw him in the campus gym, it always hyped me up to push myself as well. Always giving off that, "Push it to the limit!" vibe, or "No pain, no gain!" He started working out about the time I entered high school. So, when I met Jeff, he was already well into the termination phase. I am on the cusps of reaching that phase and still trending along with the never-ending maintenance phase. Here's the catch, I already know I am in the termination phase. And I have been for years now. There is no part of me, not even 1%, heading back to old unhealthy habits. One of the main reasons why I wanted to study nutrition was because of this reason, and also being diagnosed with Celiac Disease about a decade ago.

If you guys haven't heard of the craze by now, then you must be living under a rock. The gluten-free era is finally upon us, and I could not be happier. Ten years ago, living with the Celiac disease was certainly a headache, literally. I say literally because this is one of the main symptoms I had, which led to testing. Sure enough, a blood test came back positive for Celiac Disease, and I was to adapt to an entirely gluten-free diet immediately. I never even heard of the word "gluten" before, so how the hell am I supposed to avoid it? Well, before I knew it, going gluten-free wasn't all that bad. I started losing that baby fat faster than normal and started to embrace healthier lifestyle habits. I remember just about every day in high school, and I would always get a grilled chicken salad with red wine vinegar and olive oil. Let me tell you this, and it was one of the most significant investments someone could make. Before I knew it, exercise started and got more intense as time went on. One of my good friends, Nino, got me accustomed to the gym and showed me his day-to-day routines. Before I knew it, the gym was put into my daily routine too. No longer a place I went to work out, but a place to be happy and feel good while doing it.

Before I knew it, I was in my senior year of high school and needed a path to choose for my college career. One day, I was in the car with my dad talking about the subject. He was shooting a bunch of majors that interested me, but none seem to come at me like a bitch slap from an ex-girlfriend.

Then he asked, "What about nutrition?" I replied, "That's it; that is the one." I applied to a college out of state to get out of the city and gather experiences many are afraid of catching. I'm glad I decided to meet new people and travel to unknown places. This is how humanity prospers as a whole. Traveling the unknown to find out who we are, and how we make ourselves fit into this mysterious world.

Seeing the world has never been easier, but at the same time, harder. You may be thinking to yourself that the harder part is because of a phone. Well maybe, but the point of this section is for you to take in the world, rather than look at it. Earth is an organism, just like us. The planet has its past, present, and future soon to come. It has a path that has many obstacles in the way, just like each of its inhabitants. We weren't just put onto a random planet in the immense universe for no particular reason. The world around us is always revolving and evolving right before our eyes. However, times have changed drastically over the past few decades. Technology is now a factor, and it plays a massive role in our lives every single day. We have the world's information in our handheld devices. You may think that the next sentence I am going to write is, "And we take it for granted," but you are wrong. Even though it is true, the real lesson is that our phones suppress our abilities to take life by the balls.

Knowledge doesn't come from a device but instead experiencing an event and moments in real life. No one has ever and will never learn a nickel of knowledge through virtual reality. Our eyes are so glued to our phones and social media accounts to the point where there are laws to stop us from looking at our phones while on the road.

A phone reminds me of a mirror. Do you ever talk to yourself when you are in front of a mirror? Of course, you do, everyone does it. Whether you're giving yourself a pep talk, or crying about a bad day, now think of this, do you ever talk to your phone? No, not to someone on the other line, but at social media or videos you are watching. Of course, you have. Wasted emotions run through you whenever you unlock your phone and dive into social media. Here, I'll let you dwell on this. If you opened up social media earlier today and saw a funny video that made you laugh out loud, would you remember it the next day? No, of course not, that's the ugly beauty of social media. It's emotionally moving at the moment, but in the long term, it's pointless.

Our brains are our organic hard drives, but there is no space limit or capacity. We could place whatever we want in there. Ha, listen to me, acting like we have a choice of what goes in and what comes out of the most cryptic organ in our body. I feel as though our bad memories seem to linger more often than we like. We think of these embarrassing or horrific moments for days on end, and countless nights leading to sleep deprivation. Shouldn't we be able to control our brains the way we like to? We discovered how to venture into someone else's brain, but why can't we venture into our own? Just figure out the problems firsthand without needing someone else's guidance. Well, I don't have the answer to that problem, and I may never find the solution. But a key component to unlocking the brain and finding the hidden codes within us is to take in every moment. Observe what is around you. Start up a conversation with a random person in a coffee shop. Pick a different way to work some days, and wake up at a different time. Bettering yourself day in and day out no matter how small the benefit may be will be worth your while in the long run.

The Power Within Synchronicity

The mind, our mind is a good thing to lose ever so often. The one thing I ask of you guys is to remember the moments of life with full observation. Without this trait, synchronicity will be hard to find. Or should I say, it will be hard for synchronicity to find you? Synchronicity is the reasoning behind everything and anything. It has no boundaries, no limits, just like the human mind. Always evolving, and taking in new knowledge. You need to take in every moment, even if it is a simple one, grasp the moment in its entirety. Synchronicity can be defined in many ways, and it depends on the type of person you are. It is kind of like looking for something that is right in front of you, which is usually the hardest thing to find. Something so simple seems to be so far to understand and comprehend in today's world. Everything today is thought upon with too much brainpower and emphasis. Tell me not every little thing stresses you out. It doesn't even have to be about your phone, maybe someone cut you in line for coffee, or someone took the machine you were about to workout on. Frustrating right? Imagine that, first thing in the morning and you are already getting screwed. I guess that you would expect some sort of apology. We apologize way too much, and usually for no reason what so ever. All of these stressors can take a toll on the body without being released once in a while. I am sorry to say it, but staring at your phone, and looking at social media is not going to release any. It may make your stress worse.

The simple act of synchronicity is all one needs. Here is a great example I whipped up. We have a young couple; a man and woman are about to be on their way to a social gathering downtown. The man and woman are both outside, and the woman is waiting for a few dresses to dry. One blue, one yellow, and one with polka dots, just like a ladybug. Once they dried, the woman asks the man, "Which dress should I wear?" The man shrugs and says, "They all look great. Now c' mon. We're going to be late." With little acknowledge to her. As he's staring down into his black screen, wouldn't you know ladybug land on the man's right shoulder out of nowhere. The woman points and shouts, "Look!" The man gently picks up the curious ladybug filled with black polka dots and says, "Well, I guess you know which dress to wear now."

Just like that, synchronicity hit them like a lightning bolt, but with clear skies. The point of synchronicity is that there is no point. The physical law of nature states that something needs a cause to affect. Everything in this world has its column in life, while synchronicity falls in its own. Somethings that we're not accustomed to are the things we need; to make life seem more meaningful, complete, and spontaneous. The rule of synchronicity is that one cannot wait for it to happen. Pretend to be in a scenario where you look at the clock, and it is 11:11. Waiting for another 12 hours to see that exact time again isn't considered synchronicity. Synchronicity will be the spontaneous moment that catches you off guard.

The ladybug example isn't a real synchronistic moment. It most could be, but this may not happen twice, not even close to three times. Signals that may be with you every day can come from a set of numbers, or one of your friends mentioning some wacky idea you were pondering upon in near the past. Another one I feel regularly is a question I'm thinking about and someone in the future answering it for me, without me even asking. Now, these types of examples are hard to find because most people won't make the connection at the time. Say, for example, you see a license plate with the same number. Well, that would be easy to catch because the numbers would be lined up right in front of you. Now, say you see this same sequence of numbers on the same day. But, will you remember the first encounter? Many of you are probably saying, "Of course," but in today's day and age, our surroundings are taken for granted. Noticing is only half the battle of synchronicity. There are many more elements for this phenomenon to work for you daily.

These elements depend on your mindset and how you guys interpret the world around you. Some starting strategies I always fall back on are as simple as just paying attention. Paying attention has always been the backbone of knowledge both physically and mentally. Just having this constant trait alone will open many doors for you to acknowledge the sudden synchronistic moments. Start recording and remembering random things in the day. Noticing at the moment is easy, but remembering that moment is harder than it sounds. Hell, some people barely remember what they had for breakfast when it comes to that upcoming evening. Becoming aware is a crucial element to your intuition of having synchronistic events. With these few traits, synchronistic moments will not only happen more in your life, but they will come to you, like wiping after you've taken a dump. It's that easy. Oh, and one more thing, you must believe. Believe, and you will have the world in the palm of your hand. Having a positive outlook and attitude gives you the upper hand and will enhance your experience in life. Making the time to practice your skill of catching synchronized moments will be difficult because many fail to have the patience. You can't go outside for a couple of hours waiting for something to happen.

Here are a couple of steps up from the ladybug synchronistic event:

"I heard a story once about a man who was having a hard time changing careers. The process was causing him a lot of stress with all the doubts, twists and turns, and unanswered questions. He'd been offered a new job, but he wasn't sure if he wanted to leave the security of his current job where he'd worked for years. One day, as he was driving to work in a state of indecisiveness, he asked himself the questions that had been churning in his mind: "Should I take the new job? Will I be happy? Am I doing the right thing?"

At that precise moment, he looked up and saw a bus drive by. The billboard on the side of the bus had a Nike advertisement with the slogan: "Just Do It!"

These are not just coincidences! I don't believe there are accidents in this intelligent universe. It is times like these that you should ask yourself: "What am I supposed to be learning or doing right now?" I feel when these synchronistic events occur; it's the inner workings of your soul showing up in your outer physical world.

Synchronicities act as signposts or mile markers, guiding and directing you, or even helping to align your personal growth. I want to encourage you to notice when you're being sent such signs, signals, people, patterns, books, articles, and so on in your life. I think of them as little gifts. These unique gifts help direct you on your path toward your goals, and help you follow your soul's guidance-all you have to do is ask, and be watchful."
-John Holland

I believe this is an inspiring, yet simple story. That is another thing I want to point out. I do not believe in coincidences. These happy accidents aren't accidents, but rather the universe talking to you to act now. Now, this right here has nothing to do with your handheld device whatsoever, and there is a reason for that. These events will not occur within a phone. They never have, and never will. So, give that constant distraction a rest, and let the universe do the talking and guiding.

Good Questions Lead to Better Questions

Who the hell wants to hear the answers anymore? Let me reword that I don't want you to close the book on me now. How about this, what if life was just the same crap every day from here on out. I bet most of you are saying, "Yea, life already is like that, smhmuck." Jeez, tough crowd. Alright, I'll try to break this shell. But c'mon, you guys have to lend a helping hand, so this knowledge I am about to bestow upon you will not be taken for granted. I can't do all the dirty work, that isn't fair. Plus, it takes the fun out reading whatever this is. Hopefully, you keep reading, and instead of looking for the answers, look deeper for the questions. Fall under my hypnotizing and mesmerizing propaganda. Don't bother to try and fight it. This book will always be staring right in your face, even when you not near the darn paperback. Why accept your fate, when you can question its entirety too? The basic blocks of knowledge have always started as a curious question or questions. Answers are the puzzle pieces while the questions are the directions. Both are as important as the other, but you decide where and when the next piece will fall into the puzzle. Kind of like a peanut butter and jelly sandwich. Who the hell wants a peanut butter sandwich? Some schmuck, that's who. And guess what, you are nowhere near that schmuck. Cause what you have done in your life up to this very moment defines who you are as a human being to others around you. People around you, including family, friends, co-workers, and even teachers, learn from you every single day. Learning how you inhabit and take control of your life. Questions are always getting blurted out of the mind, but answers are usually hard to come by.

I bet you already know what I am trying to get at here. "Answers are usually hard to come by." Come on now; you could easily compare this to the art of synchronicity. The art of finding the element that defines you as you. It is the backbone of finding yourself. It's like the old school game that is played rarely today. Oh, what's it called again? How about we play a quick game of "hide 'n' seek" for old times sake. But here's the catch. You, the seeker, aren't exactly sure what or who you are trying to find. You are just going to count to thirty seconds and go looking blind as a bat. You have to do something out of the ordinary to try and find the hider. That menacing hider is an intricate player with several tricks up its sleeve.

Now, before you start looking in trashcans and closets, look for the answer in front of you. Don't make the game too complicated; remember your competitor has the same lookout on the game. Who wants to make their lives difficult? The easier route, no matter what is always the go-to route. Just like everything else in life. Say you are coming home from work one day and you are in the midst of rush hour. You look up at the GPS for alternate routes, and you see one route is ten minutes shorter. However, you aren't familiar with this route, do you still take your chances for an extra ten minutes? I'm sure 99% of you guys would take the shorter route, and hey, who can blame you? Taking the easy route these days in anything we do is the go-to it seems. However, say the easy route won't give you the same result as the other route. You still get the task done, whatever that task may be, but the pleasure is decreased. Hmm, I am sure many of you are shaking your heads at this one. The more complicated things become, the more it becomes a hassle to us. But, what if I say: the more complicated, the better the reward. Now, all of you guys would want to take the hardest trial out there for the best reward. Yes, that may be easy to say:

But do? That is a different story that will probably never be told. Right, why the hell would we want to make life more difficult than it already is?

The way we lead our lives today is a big enough hassle, even with minimal difficulty. We make things a hundred times more difficult because of the way we think, talk, and act. It is already hard enough to keep our core sanity safe and undisturbed. Now, you are asking me to perform a task that very few ever accomplish? Count me out. Here is the key message that many are afraid to hear because it sounds too good to be true: The simple task of saying less, and asking more. In today's world, our brains are encased in a shell of assumptions and taking feedback wrongly. We all think we have the answers to everyone's problem, including our own. But we don't. Scientifically, none of us do, but this is where we learn and excel together, even if no one is on the same page. Instead of looking for the answers to problems, one must instead look at the base of the complication, which are the questions that are lying underneath the surface. A surface we aren't familiar with. The world wasn't built with masterminds, but rather deliberate conservators. "Having a conversation? Oh hell, this will be a piece of gluten-free cake!" It isn't just any conversation, and sure as hell isn't a conversation consisting of two text blocks saying,

"Heyy," or "What's up?" C'mon, how dry does that conversation starter sound? Jeez, what are we, in middle school? Get that crap out of here. But anyways, can we change our current habits of gathering answers into asking questions? Well, as I said before, with synchronicity, practice makes perfect. What I always find helpful when I ask a question is to first realize that regardless of what the answer is, remember that there is a potential range of other responses that are the answers. Pretty mindboggling stuff, but hey, you will get there. I know for myself, I am still trying. I'll tell you the truth, these days, and I would rather receive an answer of "no" rather than "yes." Boundaries get broken, and the true light of knowledge will seep in. The process overall will lead your future self into a more elusive and smarter decision-maker when it's crunch time. Your attention span will be far greater than a pea, and your focus will have you feeling on the same level of a double dose of ecstasy. Your results that you deliver and take in will be far different.

Putting together the puzzle is the easy part. The difficult part is where people are confused about what they are looking at when the puzzle is completed. That sucks right? Just when you think you have it all figured out, it still doesn't make any sense. It is like the same feeling of finishing your school career. Years of hard, and stressful work that you put timeless hours in to day and night. Now that you finally have the final piece of the puzzle (which is the diploma), life will start to fall into place with no if's, and's or but's. It doesn't work like that even though we were made to believe in that in our society. Life is a never-ending puzzle with its question and answers. However, there is no painted picture of the final, and completed masterpiece. That is for you to decide; where each corner piece goes. Our "puzzles" are never-ending, and you know what, that is a good thing. If it were to get to the end, eventually, we would all go insane. Every piece you place has a different meaning that only you understand and hold onto deeply. But for now, let's focus on the corners where your deadlocked beliefs are: The basic blocks of your life, which won't precisely focus on memories or current circumstances. I don't want you to focus on close family either, even though that is part of the base. Your perspective of life will be placed here — the way you function properly with yourself and around others.

The four corners will represent you and build off those corner pieces to reach the middle. Keep in mind, don't strive to connect the whole puzzle. The four corners are mainly the questions that you don't want to answer. The questions that you will keep picking away at, discovering new ways to ask that question. Always maneuvering the piece so that the mind will always keep you up to speed. Finding the answer to anything complicated in the beginning may be an extravagant feeling. But that feeling won't last very long. That feeling may be forgotten the next day for crying out loud. Always pondering on a difficult and borderline impossible question or acquisition is what makes our minds work, and also keeps us sane, even if it doesn't feel like it. As you go deeper into your puzzle, more memories will be made; consisting of coming up with a fantastic joke, or a regretful one-night stand. Emotions run fluidly throughout everyone's life puzzle like a roller coaster with multiple twists and turns, ups, and downs. Just follow your heart, and fate will soon be knocking.

Accident? Luck? Coincidence? Fate!

The four words that make the world go round are just about everything we all feel daily. However, all of these moments in time have a meaning, even if you don't think so. One of the most "coincidental" moments of my life happened back when I was in the third grade. I don't remember much from that time, of course, but this moment will always be remembered. So, let me ask you this, say one day you go out into your neighborhood and you run into one of your friends. This is certain a likely occurrence because you two live so close together already. Now, think about seeing them in another country. You probably never would right? Well, when my parents took me on a trip to Italy, I ran into one of my friends in my grade. Of course, my parents stirred up a conversation with her parents, and all is said was a simple "Hi!" At the time, I didn't think of anything, just a simple coincidence.

Now, let us step into the future. Me and my good friend Dom, one of my roommates, were on spring break in Daytona Beach, Florida. While having a good time at the clubs - what do you know! We run into some students from our college in New Hampshire. We met up, and they had a table full of vodka and other inexpensive booze. You may think, that is pretty likely because of what a hot spot Daytona Beach is for college spring breakers. Now, let us step one more year into the future when my father and I went to Miami Beach during spring break. What do you know, I ran into the same people. I couldn't believe it. The moral of the story is that what the hell are the chances of this happening again? Once was plenty, but two times! It is almost as though we texted each other to meet up, but no devices were involved. Is it luck? A Coincidence? Or is the universe just setting up the perfect time? Synchronistic events like this happen all the time, almost like magic, but there is no trick to it and no direction to take. It just happens, so keep your eyes open.

Before my graduation from college, I had to study for a very important test to continue on my career in nutrition. At the time, I didn't take it that seriously because I still had to complete my finals in all my other classes. This test consisted of one hundred multiple choice questions about every field of food imagined. Ranging from medical needs to food management and safety in a kitchen.

Of course, the first time around, I failed even though I was very close to passing. I felt down on my luck and headed straight for the gym to relieve some stress. Who do I see there? My main man Jeff. I told him about the test and how I came up short. He told me not to be down and instead, build on the mistakes I have made coming up to the test. Luckily for me, there was a re-test about ten days after graduation. I felt blessed to get another chance, so I started studying right away. Making decks of flashcards based on the study guide and asking roommates to help me study and test my knowledge on the subject. When it finally came to the day, I was about to puke. I was so nervous because I don't know what I would have done if I were to fail the test for a second time. I should mention that the questions were completely different from the first test, so the second time around wasn't exactly a walk in the park. However, that morning, when I walked into the classroom, I felt confident and ready to take on the beast. By the time I got to the last question, I had to pee so bad, but I still took my time on question 99. After another click, I thought to myself, "I will either be the happiest man in the world or the biggest failure yet again." After pressing that button, I got up, ran down the hallway to the nearest restroom. Opened the stall, and let out an ocean of pee, flushed, and walked to the sink to wash my shivering hands. I looked up into the mirror and started shouting, borderline screaming, "YEAH!" Flexing just about every muscle fiber in my body. I walked back into the classroom, trying to calm down as other students were still trying to finish the headache of an exam, and gave the professor a thumbs up as I walked out through the door. My friend texted me after to workout, and of course, I obliged. Take a wild guess as to who I saw at the gym. Yes, Mr. Coach Jeff. I walked over to him with a big smile, and he already knew what the news was. I gave him a huge hug, and it felt so freaking good. Fate has led me to Jeff after failure and success. You may think that this is just a simple coincidence, but to me, it means something a lot bigger.

Throughout the time while I was studying, I tried to keep distractions at a minimum. Days on end, I wouldn't even think about touching my phone. The way I did this was putting it out of sight. Once that sucker is out of sight, it is completely out of mind; unless it rings or makes some noise. Distractions come from every-where these days; rather it be noise or a rancid smell that comes from your roommate's room down the hall. But you cannot lie, with that phone of yours lingering in vision, it is almost as if the phone is talking to you, to pick it up and mindlessly look at it. Phones take away about every ounce of time we have these days. It is to the point where it is deteriorating our minds and disturbing our daily function. This distraction lingers in children with their playtime and adults with their work time. What is so important about our phones to make us procrastinate on our daily tasks? A phone's value, whether it be a flip phone or touch screen, is so great that people would do anything for it. Cracking the screen or maybe even breaking the phone all together by accident is the worst fear someone can have these days. Because without a phone who are they? Who are you? Who am I? That is the thing, though. It is easy to buy a new or refurbished phone. People don't even care what they are paying for a phone. Just having the deluxe version feels better to them, it is as though they feel more complete. We don't even need our minds to solve a problem anymore. We have the skyrocketing price of artificial intelligence to do the job for us.

The World We Now Live In

Life Through A Device

Touch $creen

Ah yes, you already know what is going to happen in this chapter. I don't want you to feel guilty, embrace what is about to come. This section is about a phone, specifically the one next to you right now, but this is only a small piece of what a giant the technology industry has become. I don't know if you guys keep up with the stock market, but the DOW industrials, Nasdaq, S&P 500, and other markets around the world keep reaching new highs every day. The reason for this is because of the growing markets not only in technology and medicine but in money. Yes, the market for the money. The demand outweighs the supply by tenfold at this point. The needs and wants Americans have, are only going up and will keep rising with all the new gadgets coming out day after day. Increasing markets means more expensive stocks, which leads to more costly commodities such as water, eggs, and even gas.

Who knows when the next market crash will be, hopefully not anytime soon, but the next one will surely be a doozy. Will all the tech stocks drop as well, or just the industrials? Is the mighty

"Bitcoin" here to stay and take over the way we spend on a day-to-day basis? Not only are we going virtual, but our money also! Either way, the dollar will come crashing down as prices for goods skyrocket. Numerous jobs will be lost and Americans will go bankrupt. Whether it be stock portfolios hitting an all-time low or houses start to go up for sale. Panic will run through every street in the nation. However, in today's day and age with the ever-growing market and blue-chip companies such as Apple, Facebook, and Amazon, it is hard to believe and foresee the markets ever crashing again. These giant conglomerates will continuously keep the markets afloat even if the stocks are going down. The internet and all of its subsidiaries have such a significant impact in our lives that the web as a whole plays a vital role in the markets, the prices of goods, and the salaries, and hourly wages we are given. You could see why the web is part of our lives day in and day out. Can this very well be the rest of our lives? Just puppets of the show always playing the sidekick of our phone. It is like the value of life grew less. Here is your new and improved American dream. Just being a "follower" and asking for "followers" to "follow" you. The never-ending cycle of humans just moving in a circle. Today, tomorrow, and probably the next day will be followed by distraction after distraction, making you lose sight of your true calling. The ludicrous spending on these devices makes us not want to put the damn things down. Now, getting away from the ridiculous price these devices come with, let's get on to other matters that worry me even more. I recently went on a trip to Florida, to attend and head-bang at a music festival with my good pal Dylan, who lives in Tampa. I boarded the flight that was heading from JFK airport to Tampa International. The flight was about two and a half hours long, and I was eager to find out who I was sitting next to. When I boarded the plane, I was only a few steps from my seat, and I saw a quaint Asian woman next to the empty seat. I had two bags, and the space for luggage above the seats was already packed, so I had to sit with the annoying baggage. Luckily, once I got to the seat, Luna, the Asian woman next to me, let me put the bags in between our feet, so we got an equal amount of room. So right off the bat, I knew she was a nice woman and wanted to have a conversation. We talked just about the whole flight, from take-off to landing.

But the main topic I want to get out to you guys is that she works in a hospital in St. Petersburg, Florida, which is about twenty minutes south of Tampa. She told me pretty moving stories that make you think, "Wow, I cannot even imagine that happening to someone I know in my life." This man from Canada was taking a vacation in Florida, and while on the golf course, he blacked out and woke up in the hospital nearby. Unfortunately, Luna received the sad news and had to tell him what his current situation was. He had three to six months to live because a tumor was found in his brain, and spread like wildfire within his cranium. His family was there by his side within ten hours, and the look of defeat filled the room. I was thinking to myself, how in the world could this happen without any warning signs in the past? Luna was baffled herself, and went on to say, "I am starting to see tumors in the brain of younger and younger people, one girl as young as 19, had a tumor in her head!" She even said to me, "I would not be surprised that these tumors are caused by phones." I am sure you guys have heard the rumors about how phone radiation can cause cancer. I know my mother is always nagging at me, "Will, take that damn phone out of your pocket!" Well, I will probably start doing this soon enough. I don't like to point any fingers, but cancer, especially in the brain, testicles, and breasts, have skyrocketed over the past couple of decades. Unhealthy lifestyle isn't the only issue anymore, but also too much exposure to radiation. This theory may still be too new to call this a causal factor, but you can't even deny it.

The Daily (every 5 minutes) Distraction

Alright, so you have been sitting for about a minute and a half doing nothing, yeah, sure pick it up. So, you have been waiting in line for a cup of coffee for approximately 15 seconds. Scratch that, even before you walk through the doors, you notice there are people in a line. What do you do? Try to start up a conversation with the last person in line? Maybe, "Oh, what beautiful weather we are having," or "which flavor are you getting?" No, you dive into your pocket or purse for your phone. The sad part is that you have no reason to go on your phone whatsoever. C'mon, let's be honest, no one has texted you or called you. It's only 7:30 in the morning. You are just doing it, so you don't feel out of place. But here's the thing, everyone in the coffee shop will be guilty of doing the same thing, just staring at a blank screen with a blank expression. As if we weren't already zombies when we woke up in the morning, but now, we have our phone to make us walk slower, and act slower.

A phone certainly has developed this world around us as we know it. I'm saying this as a good thing because we have evolved so much over the past couple of decades. However, I feel like we, as a society, take these devices for granted. What would happen if your phone and every other phone just disappeared? I wonder how the world would run that day. Even thinking from an individual perspective, how will we all act? A day like this would go down in history, "The day that went dark." People would be running and screaming down the streets, shouting, "The world is ending," or "It's a Judgment day," something ridiculous like that. Panic would course through major cities and businesses. It's only for one day. Can humanity make one day without a phone? What about a whole week? Jeez, I don't even want to fathom that thought.

I want you to think about it now. Devices such as the ones you may be holding right now while reading this sentence weren't part of your daily life about a decade ago. Our structural regimen of everyday living was contorted, and pretty much tossed away. The industrial age is starting to wither away as the informational age is upon us and rapidly growing. This new age of technology is changing the way we think, act, and prioritize our lives. Do you let your phone get in the way of your day-to-day assignments? Constantly making excuses? Ever looking at pictures on social media, contemplating if your friends and all the random people you look up are consistently happy daily? Always having fun, and not a care in the world. It is all very deceiving if you ask me. I feel like everybody you meet now has social media where they post about their life, the stuff that matters and will attract people's attention, an attention that comes from you and me. However, I want to know what is on the other side of that coin. In other words, this may only be half of the story about what these people do and how they feel. I am not saying that everyone that logs into social media daily is depressed or feels social anxiety more frequently, but one can still connect the dots. It is even easier to tell when you know the person well.

Say, for example, one of your good friends is going through family trouble and feels emotional pain. A pain so deep that you feel it as well whenever you are with him or her. They give off that uninviting vibe that makes you feel uncomfortable. However, those who do not know your friend as well as you do may not know what the trouble your friend is going through simply because she hides these emotions through the joyful posts, she puts on social media. She is patiently waiting for likes, views, and attention. I believe this makes your friend's anxiety worse, not on the outside where anyone can notice it, but on the inside. The situation gets bottled up and thrown deeper into the abyss of self. Now, to all of her "friends" on social media, you may know this person through pictures and how happy she may look, but on the other side of that screen, there is another face/body/mind you don't even know exists. You know that saying, "You can't judge a book by its cover?" Well, I'm sure you have, unless you have been living under a rock, you can apply that saying here. Deception has never looked so good until now. Those who crave that attention over social media are slowly, but inevitably drowning in a sea of self. The only way is to somehow change a person internally rather than externally. Think of it like this; we suffer not from the events in our lives, but our judgment about them.

I could not blame those who want some approval, even if it means going through social media. That short but sweet feeling of admiration can make someone feel resilient and confident. Whenever I post something on social media, it is not for all of my friends to look at, but it's what I want them to see. Here, let me explain it in simpler terms. I want people to know that I am enjoying and loving life every single day. I can display that reality through social media. That praise can get me through some rough times. However, those turbulent times still linger, if not become even stronger.

I had no idea that you spell "Secrecy" without a "t." You learn something new every day. Anyways, we keep our secrets only for those we are incredibly close with and trust our own lives with. Whether it be a new and innovative business idea or a weird-looking bump by the genitals. If this were to get out to the public, that person wouldn't know what to do. Keeping secrets like this is very necessary. Only you and those you trust should know what's behind closed doors. However, I feel as though today, we have secrets for just about everything. And what's cool about secrets these days is that we can keep them locked in a place that only you have access too! A tangible place, known as a storage device on your phone. I always wondered why people make sure to point their phone in a direction where no else can look at the screen. Whether it be the person looking at something or typing something, nobody can see or else, their reputation will be in jeopardy. What are you hiding on that phone that is so important?

One thing that always freaks me out is when I am with a girl on a date, and just about the whole time, she is on her phone — of course, tilting it on its side so that I can't even glance at it. Give me a break, and there is no way that phone of yours is more charming than me. I've always known what she was doing on that thing, but I didn't want to admit it. She put it out there in plain sight pretty much saying "I DON'T WANT TO BE WITH YOU!!" Jeez, Louise, I wonder what she did when we weren't together. You know what, I don't even want to think about it. What I am trying to get at is that people keep secrets because they don't want to face reality. I mentioned this earlier; people hide behind their phones and the social media craze. People aren't themselves anymore.

Our phone addictions can be compared to the same kind of addiction that comes from drugs and alcohol. It isn't exactly the material aspect, but more of our behavior. We don't just use a phone to get something done; we also use it to lift our dreary moods. But here's the thing, it could take longer and longer for someone to uplift their mood from a phone if that person is already constantly on it. I believe a phone already acts a stressor when it is on, and ready to be used, but now, say the battery is running low and on the brink of dying. Anxiety will set in for the person because the withdrawal stage will begin. It's like that nostalgic episode of Spongebob Squarepants when he is trapped in Sandy's tank with no water to breath. It is crazy to believe that not one of us can make it through a day without a phone. The stress of not having your phone on you or it not being in sight can raise panic. Is there a way to ease our way out of all of our phone addictions?

You are probably thinking to yourself that every lesson of the Transtheoretical Model pretty much resembles the same teachings for decades now. Even though it is similar, it is far different because the lessons I learned throughout Jeff's class always made me think of the model. So, I never forgot it, and never will. Applying it to yourself every day is the main challenge, it isn't just a one-time thing, and you have it down. Practice makes perfect in this world. Going through life never has one direct path to happiness and bliss. In life, you must always ask yourself, and other open-ended questions. The "what, how, and why" questions will expand your knowledge more significant than the "can, could, should and did" questions in life. You can't go deep enough with close-ended questions. Answers that only consist of "yes and no" aren't nearly enough to learn about something or someone.

Alright, now, getting away from the facts that you, the reader, already know. I want to know something from you. Here is one simple question. How much time a day do you spend on your mobile device or computer? How often are you looking at funny videos or updating your status on social media? Don't worry; I'll wait, and don't be vexed, to tell the truth, because, we all (including myself), fall victim to the naive act. Many sources say that the average time a person spends on their phone daily is a total of 4 hours. 4 Hours! No wonder why people make excuses, or say, "I do not have time for that," or, "I'll start working out tomorrow or next week." You may be saying to yourself, "There is no way I am on the phone for that long every day!" Now tell me this, what is the first thing you do when you wake up? And the last thing you do before you go to bed? I'm guessing the question has already been answered. It's heart-breaking to see how much of that time is wasted that could be put to something more engaging, such as walking the dog, planting a tomato seed, or writing a book.

Not only are people looking at their phones for long and multiple times a day, but are also looking at them at the most inconvenient times and places as well; such as at the family dinner table or even worse while driving a car. "Texting and driving" wasn't a thing 15 years ago. And within these 15 years, millions of damage costs and many lives have been taken to the senseless act. I believe that texting and driving are way more dangerous than drunk driving. Now, don't get me wrong, I think drinking and driving are risky, but at least, that person will be paying attention to the road. People scrolling through social media while on the road can be described as an unmanned bulldozer rolling through a shopping mall's parking lot. Scary to think, right? Think about how many times you heard, "Don't text and drive." I bet you can't even count that high. However, I do respect all the danger signs and education that has been put into this careless act.

So, we got people on their phones at the beginning of the day, in the car, at the dinner table, and right before bed. I wonder when else people go on their phone. Oh, and by the way, that little list of when people are on their phone can already accumulate to two hours. How about on break at work? There is no way you are holding a newspaper. You brought a phone charger to work so that you could use it, fully charged at break time. Boom, another 30 minutes, wasted. Are you watching TV? I wonder what else you are doing at the same time. Commercial break? Do I even need to say it? Hours pack on just like that, without you even realizing it. Now that I mention it, I haven't paid much attention to that dusty old thing in a while. I watch sports here and there, maybe the morning news, but that is it. Damn, now that I'm thinking about it, I think I'm going to find a good show to get lost in.

TV, Now a Thing of the Past, and Hell Do I Miss It

I'm sure you are questioning this one, and hey, I understand. Who doesn't love television? After a long day at the office or hearing your girlfriend bitch all day, I bet all you want to do is sit back, relax, and watch some mind-numbing crap. Alright, I get it, your life still revolves around the TV. But let me ask you this, do you also stare at your phone while the TV is on? You probably can't even answer that question because you forgot what you just read. Go ahead, and reread it.

During the early stages of my life, I would live and breathe the ol' teletub whether it was watching a mind-numbing cartoon or becoming a zombie while playing a videogame. The phone age took some of that part of me away, but the main cause was going away to college. Going away to college was an experience like no other. I remember my first day moving in, moving away from family and friends; I was scared to death. Everything was different; from the environment to the way people talked. Turning to a whole new chapter made me a whole different person. The person that was addicted to videogames and watching endless hours of TV was starting to fade away. Going on random and countless adventures as a college student sculpted me into the person I am today. Being on your phone or watching television sure as hell won't have the same effects.

My body and mind were surely developing at high speeds when I was away at college. Meeting new people and trying new things brings out the best character in a person. I was also blessed to have a spectacular recreational center on campus. I was thinking to myself before while writing the earlier chapters, "How could I not talk about the gym? Where the mind surely exceeds." You might be wondering, "Where the mind exceeds?" I am 100% with you on this one. The gym isn't just a place I go to workout and be healthy. It is part of my daily routine. A habit I cherish and enjoy very much. It is at the point now to not even look good but to feel good and release stress. Alright, I get it, you don't want to hear me blabbing on and on about the gym, but I had to put it somewhere, and it seems to flow pretty well into the next paragraph.

The fatter America is upon us, and it has been for a few decades now. Where obesity and overweight rates have skyrocketed at an exuberant speed, people can't even blame the food anymore because we all know the dangers that come from refined and fast foods. Hundreds of deadly diseases can be listed from foods that promote obesity. However, we have known this fact for decades now as well. People tell us everywhere what the best diet is to lose weight and cut fat. We get it; everyone is a damn nutritionist. People also ask us to get up off the couch. I don't know about you guys, but for me, it is rare to have a completely off day, where I just sit on the couch all day and don't leave the house. I am not saying it is a bad thing to have those days, but I can't do it, not my cup of tea.

Alright, here we go, scenario time. I want you to imagine yourself in the gym. You are having a good time lifting weights or running on a treadmill. Now, look to your left, and now your right. What did you see? Did you saw other workout enthusiasts, right? Eh, maybe, you may see other there beside you, but I bet some of those "workout enthusiasts" were either glued to their phones or taking pictures of themselves to document that he or she went to the gym. If you think about it, you could add about another 20 minutes to that total time on the phone in a day.

Now getting back to the matter at hand. Our TV's grow dustier as our phone screens get more smudged with our fingertips. Even before phones, newspapers got attention even with the daily news on every single hour of the day. But now, with the constantly updating news at our fingertips, there is no need to read a paperback. That crisp feeling of holding a newspaper is crumbling away as our phones become the new source of what is going on in the world.

Speaking of newspapers, have you noticed that social media has mostly been making the headlines? Not the news, but the fake news that the people care about. I saw something pretty funny today concerning Donald Trump. Alright, don't get me wrong, I will not and never write about politics. But, his Twitter account made page two of the paper because it got deleted for eleven minutes. Some employer on his last day had one last hurrah. How could something like this be more important than what is going on in the world, our world? People like me and hopefully you, can't give a crap less about social media, especially making news headlines. The newspaper lacking the drive to make better headlines is the reason why the industry is tanking. People are too interested in short articles on the phone rather than two or three pagers in the daily newspaper.

Well, now that we are deep into our innovative world, we could see the changes television is taking to keep up whether it be futuristic tv shows, such as 'Futrama' or sports trying to incorporate every visual effect possible to try and attract viewers. I must say watching sports in high definition sure is a pleasure. Watching play by play in excellent quality assures you won't miss a second of the action. It's almost as if it's better to watch sports on the couch, rather than in the bleachers with players yards from your seats. The atmosphere isn't the same, but the comfort of being on your couch? Oh, wowzers, that is a tough one.

Say it's the Houston Astros playing the L.A. Dodgers in California for example. This would never happen since they're in different leagues unless they somehow both reach the world- series, but anyways say you are watching the game on your smelly old couch drinking a skunked beer. The beer is cold so you couldn't care less, your phone is on the table out of reach, and out of mind. As you're watching, you notice that the camera that should be focused in on the pitcher keeps filming the crowd as well. And as you take a closer look at the crowd, you notice that not many of them are even paying attention to the game. You squint a bit and realize that the attendees are looking into their laps specifically at their phone. When you finally realize this, you blurt out all the skunked beer in your mouth. You are baffled by the fact that everyone, every gender, and every age would rather pay attention to their phone than the game they bought tickets to see and cheer on.

Speaking of sports and all its history, you may notice that kids are not as interested in the "old fashion" of passing the time. It is a real shame to see baseball, basketball, and football go down the tubes. Before you know it, the apps on phones will become the sport that every kid wants to participate in. Hell, it already is. Playgrounds become emptier and eerier as time flows on.

So now, you may realize that you could try to escape your phone once in a while, but someway and somehow it comes right back at you like a boomerang. You throw it away, and it somehow rewinds in your hands. Everywhere you go, a phone is in the midst. That phone doesn't just control the person, and it controls everything around that person as well. It has the Antenna power to control all life and all flow of work with a single click of a button. We now live in a society where the phone may as well have a mind of its own. Hell, it already has the mechanic brainpower that immediately dominates our own. All it needs is legs, arms, and whooping punch line. Our phone doesn't display who are we in life and what we show, it is us. Our phone and life on there is the spitting image of us, and all of our wicked gloriousness.

Reference to *Black Mirror*

There is currently a show out there called *Black Mirror*. This show is futuristic sci-fi based, set up into separate episodes. Every episode is an entirely different story. Episodes show the dark side of technology and how it is taking over the world. Tech didn't come with the dark side; us humans created that side for them. Remember how I mentioned earlier that phone is like a mirror? Well, now you know why the show is called "*Black Mirror*." The show conveys how all of humanity can easily fall prey to our own devices, and how they take control of our lives every single day. If you are enjoying this short novel so far, you will like the show. It blows this crap out of the water with its wicked ideas each episode. As this chapter goes, I am not even going to mention all of the references. Because I know you will pick up the similarities and connect the dots.

The real reason why I brought this show up is to dive into one of the episodes, "Nosedive." With many episodes that have already aired, this one is the one that most compares to this novel. Now there certainly will be some spoilers within this chapter. I am not going to say every single part, but I will write down some important and moving quotes and several scenes that I thought fit right in.

The main character, Lacie Pound, is first seen taking a jog with her phone in her hand. She is on some social media application where the person can rate others through five stars on pictures they have posted. However, in this dystopian society, the ratings you get is the actual reputation you have. For example, around 4.5 rating and up your considered high class, a rating of 3-4.5 is middle class, and any rating below three is considered the lower class. I should also note that in this society, people can look at someone's rating without even looking at their phone. High tech lenses are embedded in the retinas of their eyes, and it will let them unlock social media and view anyone's virtual profiles.

As Lacie is about to head to work, she encounters her brother playing videogames. Her brother is probably one of the most influential characters in this episode with only two scenes. However, the creator of the show makes it seem that he is one of the most irrelevant characters. At work and just outside in general, everyone has their phone in their hand, rating everyone around them.

Rating every time somebody makes a subtle movement it seemed. Lacie is currently at around a 4.2 rating, and she desperately wants to get to a 4.5 rating so she could move into a more luxurious area within four weeks. While sipping on her latte, she goes through social media and is glued to this woman, Naomi, who is a 4.8 rating. The watcher can quickly tell that Lacie is extremely jealous of Naomi's perfect life.

As the episode goes on, Lacie posts a picture of this doll named Mr. Rags, and she gets a call from the woman of her dreams that she stalks and envies. The watchers now realize that Lacie and this woman were "friends" long ago. The call concerns Lacie being the bridesmaid at her upcoming wedding next month, with her "perfect husband." If you ever watched the blue mountain state, you will like this one a lot. The main reason why Lacie agreed to this out of nowhere is for her social media rating to increase to reach that high-class level.

She works on her speech and practices it constantly throughout the episode. And the very first line, "In this world, we are all caught up into our drama, it is easy to forget what matters," I find so hypocritical yet captivating. As she is about to leave for the wedding, her brother asks, "What are you hoping for, 4.3, 4.4? I miss the normal you, before this obsession and we had conversations, remember? This whole ranking thing, comparing yourself to those who pretend to be happy. Your friend Naomi, I bet is suicidal on the inside." She ignores the whole ordeal her brother lays on her and she gets into her taxi. This right here, I believe is the most eye-opening scene in the episode. Her brother tries to remind her who she once was before social media. She didn't bother listening, and she went her way to the airport.

As the episode goes on, Lacie runs into trouble that slows down the process of getting to the wedding, and her rating drops dramatically, to about a 2.5. I won't go into detail, but eventually, she got to hitch a ride with a low rated trucker. The trucker the other very influential character in the show has a very heartwarming, but sad story. I won't ruin it for you, but all you have to know is that after that incident in her life, she dropped the whole social media thing and just lived without a care in the world. However, Lacie is still determined to get her ratings up even with very critical advice.

This all changed when Naomi called Lacie and told her not to come to the wedding anymore because of her low rating. Lacie was still determined to make the wedding no matter what. She eventually gets to the resort and sneaks past the guards. Once she arrives at the scene, she starts to give her speech. The speech goes terribly, and all the attendees at the wedding are appalled and give her a bad rating. As she is about to get hauled out of the wedding she takes out Mr. Rags and puts a knife to the doll's throat. This didn't change anything, and Lacie is lifted off her heels to the exit. She was escorted to a jail cell where she meets a man across the hall from her in the other cells. They scream and shout f-bombs at each other, and the screen goes black.

It is evident at the end; Lacie doesn't care what anyone thinks of her anymore. It shows at the wedding and in the jail cell. She says whatever she wants and could care less about her rating. Everything seems to click in this episode for me, except for one crucial aspect. I have a theory, but not exactly sure what to think of it. That one aspect is the relevance of Mr. Rags. Mr. Rags is a teddy bear that Lacie and Naomi made when they were kids. My theory is that "Mr. Rags" is a reference to the good old times when people weren't rated on social media when things were simple. I don't understand why she put up a knife to a stuffed doll. Trying to connect the dots still didn't make any sense to me. It is still a mystery in my book; maybe you guys could put the pieces together.

In my eyes, the moral of the story shows how society is now run through a screen rather than your self. The images people see of you are far more important than you, yourself. After watching this episode several times, you start to see every little detail more and more. Every detail counts in this episode from Mr. Rags to Lacie's maid of honor speech. It's hard to believe that people like Lacie and Naomi exist in this world. But, it is the sad truth. All of their traits that are similar and different have a direct correlation with us, and how we view the world, through lenses. Phones control you and the ones around you. Love is no longer in the air, but within our phones, somewhere emerged into those little, tiny, insignificant circuits, which brings me to my next segment of building brittle relationships.

Romance, with a phone?

Hey, if you made it this far through the novel pat your self on the back. Many do not have the attention span or the patience to make it this far through a book anymore, no matter the length. Imagine if we knew how many pages and chapters a boy/girl relationship had? And if we did, would be able to keep up with the pace of it and stick to the terms and text? If we knew this, it could go either way. It could either be a drag of a book or an enlightening experience with many learning curves along the way. However, with social media now running our lives and manipulating the way we act, can it affect our soul mate as well? If you have kept up with me thus far, the answer should be popping up in your head right now. One can either weaken or strengthen a romantic relationship with social media these days. It is one hundred percent true. It could either make or break a relationship.

Not only do the apps on phones affect people individually, but also relationships. You already know what I am talking about. Social media has the lingo, or should I say "#lingo." No, I am not going to laugh out loud in the next sentence because this is a serious matter. One of my co-workers said one day that one of her girlfriends was getting a bit paranoid about her boyfriend. I asked, "Why is that? Is he not taking her out to expensive dinners or buying her new shoes?" I asked jokingly. She replied, "He is not posting pictures of her on social media." I almost burst into tears because of how ridiculous that sounded. I calmed myself quickly and replied, "Why would she be getting upset about that?" My co-worker gave me a shrug, and we went on with our daily assignments. An interesting point that comes to mind once again is that couples and relationships never had to deal with social media complications back in the 2000s.

For example, if a boy who has a girlfriend were to like a different girl's picture on social media, the girlfriend may take that the wrong way. Almost as if the boy is giving the cold shoulder. It's like comparing that to when a girl catches the boy checking out another girl while they are out together on a date, you know in the real world outside your living room. I don't even want to know the conversation they had, must have been a lot of capital letters and mad face emojis. It is almost as if the girl would want to be posted on social media, rather than receive a bouquet. If it gets too intense, the person may block the other on social media and decide to never talk to them ever again.

Another little reference to *Black Mirror* is the subject of blocking someone, but not just on social media, but in real life. Whenever you see that person you blocked, all you will see is an outlined shape of the person you used to know. It's hard to explain, and I suggest watching the episode to see what I am talking about. Anyways, imagine having a switch like that. Do you think life would be easier to erase someone from your life? I bet many of you are saying, "of course," and I don't blame you, it is like a superpower, kind of like going invisible, but putting that curse on someone else forever. Your memory banks will be depleted completely of all the memories you ever had with that person. It will be like a blank time.

Like keeping a relationship together isn't hard enough, but now, the barrier of social media suppresses many relationships. You would think that social media encourages relationship building, but it instead can tear it down in a matter of one post, or none at all. The scary thing is that what you do on social media on your phone can and will reflect into your actual life, where things will get messy. Social media causes anxiety when it is looked at constantly. Think about it; I'm sure you have ex-boyfriend(s) or ex-girlfriend(s), who may be posting about someone new. Does it stress you out or weigh on your mind a bit? Now wait for a second, don't all you guys "unfriend" or "unfollow" your ex's? Is that not the code to follow? You may have, but c'mon now, don't lie to yourself and say you don't go snooping around once in a blue moon to see what he or she is up too. Hell, I know I do, and let me tell you this, I wish I didn't. It is another stressor in life no one has room for. Life sucks enough already; no one wants more negativity to go along with it. Instead of being downtrodden, be happy for them. That's the Jeff side coming out of me, not just looking at the brighter side of things, but grasping that side and keeping it at the forefront of the mind.

People in love with their phone may as well marry themselves. What do you think that person does on their phone day in and day out? Either on social media, snap chatting, or taking a selfie. With the main goal of bragging about their day or how good they look with a specific filter. I like to call it "Selfie Syndrome." This syndrome is contagious with just the sight of a phone, hearing that click will attract eyes. You aren't doing anyone any good with that thing out in plain sight. People who are constantly on the phone are obsessed with themselves, especially those taking photos of themselves. Narcissism has set in where we could not go a day without taking a self-portrait of ourselves and sharing it on social media for others to see. For crying out loud, technology is making the problem even worse, with the selfie stick to a get a more aerial view of us, not to mention a drone taking a picture of ourselves. The problem seems to be getting worse and worse.

These days it seems we have more of an intimate relationship with our phones rather than our friends and family. You can't deny it; your phone is basically with you the whole day every day. Strapped onto you, not like a belt, but more of a scar or a stain. I bet you and your phone sleep together, eat together, and maybe even take showers together if you have that nifty waterproof case. Now, you see that neither the man nor the woman is the problem. The problem lies within a phone and all of its addictive features, features that interfere with day-to-day necessities, such as listening, or simple communication from one person to another.

Communication and Concentration

Talking and conversing has undoubtedly changed over the recent years. Before, we had letters, and now, we have emails where a message could be sent in a couple of seconds, instead of a couple of days. We could text, make a phone call, and even Facetime with someone else in virtual reality. It is all quite remarkable if you ask me. Never would I have thought as a kid to be able to text my friend to have a play date, let alone go face to face with someone right on the couch through a device. However, as you noticed throughout reading so far, there is always a darker side to this ever-evolving technology phase. Do you know that feeling of just being alone with no one bothering you? Well, if you do that feeling is rare. With our phones at our side and constant worries, it is hard to get away from the world and escape into your own.

This brain-draining form of communication is now the culture we live in and try to strive in. What I mean by, "brain draining" is the direct correlation of concentration we have these days. Keeping up a conversion for more than five minutes these days is like pulling teeth. However, communication on the phone could last hours maybe even days. The constant need to talk to someone through a phone is a must in just about everyone's day. However, when it comes to face to face, a different side comes out of us, the side known as true reality. Conversations we have through our phone are nothing compared to the conversations we have in real life with another human being two feet away. The main difference I see with conversations is that the ones face to face are the ones remembered more than the one's device to device.

I am not sure if you guys ever heard of the app Snapchat, but it has hit the youngest generation like a smashed softball between the eyes. So, this app lets the users send pictures to one another, and the picture deletes itself after ten seconds. However, sending nudes is still risky since the person on the other side of the screen, can "screenshot" the photo for potential future blackmail. I don't recommend it. But the reason why I brought this application up is because of one special feature it brings. The feature is called "snap streaks," and these suckers keep a daily count of snaps sent to specific people. Here's the catch, to keep this streak going at least one snap has to be sent back and forth between two people.

So, right next to the person's name, there is a number that increases day by day, if one is to keep the streak going. I know some friends that have hit a hundred. Just think of that one hundred. That is more than three months of going on the app at least once a day. It's not even a habit of opening the app at this point; it's more of a reflex. What is even eerier is that there is a sand timer by the person's name if time is running out. So the person better hurry and send a snap so they could continue the streak. I got to forty-five days in a row one time. Time was running out, so I even sent her a text to send one back, but before I knew it, the number disappeared. I was thinking to myself afterward, "Who the hell cares?" It is a silly thing to even ponder or stress about.

However, people rather have those long and "meaningful" conversations over the phone because of convenience. Just hearing a voice isn't enough for dramatic effect, for knowledge to be spread, and to get the point across. People don't want to go through the need to concentrate on what the person is saying to them. Even worse, they are afraid to approach the person in real life, so he or she will say what's on their mind through text. Never mind giving an actual phone call. Kind of like a person breaking up a relationship over a phone. How pathetic does that sound? You would think that the person would have the courage and decency to break the news in person, but instead, they will scurry behind a phone.

Alright from here on out, I won't mention the word "relationship" ever again, because I am sure you are getting sick of it at this point. So, with these phones held just about every moment of the day, do you think it is like a gateway to get out of troubling situations? Well, you can certainly run from the problem, but that problem will eventually catch up to you, and it will strike with a cruel vengeance. Conversations have changed with everyone around us, even our best of best friends. Seeing someone you know and trying to spur up a discussion is harder than you think. Of course, you will get comfortable, but why doesn't the sense come quicker to you? That mask of a phone isn't only blocking your view, but the people around you as well.

Conversing has never been easier, but harder at the same time. I feel like a lot of people these days are intimidated by society outside of their phone. I'm not saying they are afraid of the sun, but more towards approaching strangers. That's another thing why in the world do we call other people "strangers." That certainly wasn't a thing a couple of decades ago. I remember back in the day; I was always excited to hear the doorbell ring. Now, when

I hear that sucker ring, I run for the hills and usually ignore it. There is no need for those ear banging rings no more because of our phones. If the person isn't saved in your phone as a contact, you will have no interest in meeting the person, or for that person to get to know you. Don't even get me started with job interviews, in which the employee has to show up and to talk in full sentences and get rid of the entire text lingo.

Going off another 'C' word that took on a whole new meaning is the simplicity of convenience. We all crave convenience throughout life, whether it be a class getting canceled or an empty line at the nearby coffee place. However, that feeling of having a convenient moment is starting to fade away. Because now with a simple click, we could have anything hand-delivered to us without us even moving from the couch. Talk about convenience and how it has changed the world not only inside of us but also outside of us dramatically. Stores and malls are starting to go out of business because of lack of customers. You would think the opposite with an ever-growing population. However, phones allow us to purchase whatever we need at our fingertips. For example, Amazon, the dominant online retailer, sells just about anything you can think of. I have to admit it is very convenient but is putting great businesses out of the league, especially the retail department. Even buying groceries can be done through a phone. Not only is the business sector moving out at a slower rate now, but we are as well. Having that convenience makes us not want to leave the couch even more.

The food that we all eat used to be the major reason why Americans were getting so heavy and lazy. Yes, believe it or not, but the food that you eat can make you lazy and fatigued even more. Now with shopping online, we don't even have to leave our house to get something. Everything is there and can be at your doorstep in a matter of days, ranging from a new barbeque to a few pairs of double-A batteries.

People need to open their eyes to something more soothing and inviting, instead of shopping online. Going out on adventures to new stores to buy new and needless crap has never been more popular than it has been today. And you know what, I respect it. People aren't as afraid to spend. Whether it is new sports cars or some crazy color expensive light bulbs, people will buy it. I mean this could be a bad thing, and I mean a terrible thing, but hey, whatever makes you happy.

Convenience has led to piss poor competition because the stores on the streets and in parking lots come nowhere close to the marketing that the online world holds. Who would have thought that the

American dream could be accomplished from your basement or garage? It is crazy to think that all of these social media and tech giants were made up and crafted in spaces like these. For example, Facebook was founded in a dormitory room at Harvard. Amazon made by two intelligent individuals in a garage. Now, here is the tough question, do you consider these types of people heroes changing the world for the better, or money savaging demons changing the world for the worse? It may be hard to tell the difference, but for now, I'll put these people in between those categories. There are countless pros and cons to social media and the convenience of shopping online. But from here on out, let's emphasize the cons.

Social (Media) Cues

Social media has probably been around for a solid decade at this point. Maybe a bit longer, but it has undoubtedly grown at an exponential rate to just about half the world's population logging into social media every day. It has made its mark in today's society as the trademark as virtual life and a virtual personality. Do you know exactly what a social cue is? Well, let me tell you, it is quite simple actually. It is a verbal or non-verbal hint. Pretty much saying if you are right or wrong. Say, for example, you stop at a stop sign and see a car at the stop sign to your right. The right gets the go-ahead. Is that a rule? No silly, it's an unwritten rule or social cue. Here's another fun but a simple one. Say you are at an ice cream shop and ask for one sample spoon of a flavor. The customers behind will respect that. However, you go in for not only one more sample spoon but several. Now you are stepping the boundary of those unwritten rules, and people are belittling you right then and there. Oh wait – it's only in their heads, but still, you know what I mean.

What I am trying to get out of this is that social cues are now being used on social media. You know what, it makes sense. Say if someone is asking for too much attention by posting pictures every day. He or she will get called out for it. It can be verbally, non-verbally, or over text. The list can go on and on about these made-up rules. I can spit out a handful right now, but why spit when I could just put it out in writing? Alright, here we go, my eight golden unwritten rules of social media:

1. No constant pictures of your pets. (Yes, we know they are cute).
2. No one cares about your women crush Wednesday (WCW) or man crush Monday (MCM), limit that crap to every three months.
3. No constant reminder of how successful of a person you are.
4. No constant reminder of how happy of a person you are.
5. No constant reminder of how depressed of a person you are.
6. No more than two pictures when you are on vacation, we get it, you are having a fun ass time.
7. NEVER post more than one picture a day. Jeez, you're asking for a beat down now.
8. Avoid politics at all costs.

I'm sure most of you can agree with at least half, if not all of my unwritten rules of social media. You probably feel the same way I feel when someone posts constantly about their pet turtle or snake. It gets old after a while, and by that I mean it gets old after the first picture posted. Now, on one side of the coin is us searching up this ridiculous crap regularly. Mixed emotions flow while scrolling down on a screen. All of these emotions can range from jealousy to anger. Probably more along the lines of jealousy. Let me tell you, that jealousy will linger days on end not just in you and me, but in everybody. Every age group and every gender, tell me not. Say you are home alone one Saturday night, you know, doing your thing where you scroll through social media. You see about every picture is something about getting drunk or having a good time. Now, this will make anyone go nuts, including myself. That is why I try to steer clear from social media on a weekend night. Do you know that saying FOMO? Well, it is the "fear of missing out," one of the worst fears out there in today's modern world. Being left out of something these days is like a bitch slap that leaves your face bloodshot red for days on end. It's the same feeling for those who missed out on Bitcoin.

Here's the hard part. You try to get yourself to not look on social media on a pathetic night, but you eventually break. You see everyone on there having the time of their lives, while you are at home, with your ass planted on the couch. Now see, there is nothing wrong with having nights like this where all you want to do is relax and take a step back. But these nights the eagerness of going out and having a good time is constantly on the mind because for the one you want to have a good time, and, of course, the most crucial part is for you to share those good moments with your so-called "friends" on social media. So they could see it and know that you are out and about living a fun and fabulous life. Pretty crazy if you ask me, but the truth is that we crave that attention and jealousy from others because it makes us feel influential and known. This is how this generation thinks, and the one after them and so on. It is almost as if this is the new social cue we all blindly follow, and the inner craziness comes shortly after.

The Dreaded "P" Words

I am sure by just looking at the forsaken chapter title; one hundred percent of you can already tell what this chapter is about. I don't even want to say it because I hate the words, their meanings, and origins too. But here is the thing, we all deal with it, and no it isn't a crime, it makes us human. And being human is something never to forget. Both of these words are set in stone in our lives day in and day out. It is hard to live with them and also hard to live without them. And that is okay, but when we go into the extremes of these words, we not only lose focus, but we also lose our minds and self-interests. These words are similar, but completely different when it comes to their definitions. The technology age has certainly put these words to the extreme, but it isn't the thing we should put all the blame on. Unfortunately, in the next few paragraphs, I will have to spell out these words so that I could get my point across. This may hurt just a little bit.

The first word is *Procrastination.* Yes, this word, believe me, I tried over and over again to erase this word from my brain bank. This word is always lingering like an ugly cloud over everyone's head. Procrastination can come in all shapes and sizes, whether it is dealing with relationship goals or as big as a business trying to achieve the quarterly earnings quota. The sad thing about this word is that we beat our selves up over the fact that we do it every single day. Let me tell you this, what is the one thing we all do when we are procrastinating on an assignment right in front of us? That's right, we pick up that darn thing on our desk and start mindlessly scrolling through social media. It makes us human to put things off and constantly think about it daily. I recently listened to *The 5 Second Rule,* by Mel Robbins. The book is a genuinely eye-opening listen on just getting out of bed and being productive. In this chapter, she talks about procrastination. The main thing I got out of the chapter is the reason why we procrastinate. I'm sure many of you are saying because we don't want to start the workload, or are just too lazy. But here is the thing; you don't procrastinate for that reason at all. The reason why you do what you do is that you don't want to have that feeling of stress. Getting started is the hardest part. Listen up, and I'll show an example.

Say you have a family and are ready to take the next step in building the family. And by that, I mean buy your own family house. You now have the money, and the support you need to put down a mortgage on your new dream home. You may have the resources, but do you have the right mindset to start looking and making offers? You may say, "Yes, of course I do!" Alright, then start looking! "Eh, maybe tomorrow, I have a busy day today." Next day arrives. "Alright, are you ready to start the research?" "Oh, today I forgot, I have a dentist appointment in the morning, and that was the time I was going to start researching." I might start slapping. Excuses after excuses is what makes the world go round, isn't that right? Saying "Oh, I am only human;" can only go so far. Making a mistake here and there and then learning from that slip-up is the right action to take. And here is the thing, that word procrastination, isn't real. We humans made that crap up because you know what it is right? C'mon, you have to know by now, it is a silly excuse. An excuse we use every day, every morning, every afternoon, and every evening. Do I need to say anymore? Just listen to Mel Robbins to get a better gist. What she says during that chapter is eye-opening. "Don't beat yourself up over it, and for crying out loud, just start."

The next word, I will be speaking of is probably the exact opposite of procrastination, but we still linger on this word like no tomorrow. We are all guilty of trying to be a *Perfectionist*, whether it be in work or our relationships with others. The sad thing is that we beat ourselves up if we make a mistake and that right there is a red flag. We all want the perfect life with no flaws, no problems, and no people trying to bring us down. People, like you and I forget that our problems define who we are, and how we fit into this world. All of us are unique in a way that only we can describe. No one else can describe your feelings except you and you alone. Don't let people tell you how the future will turn out. You set your path and you are put into the driver's seat. However, don't feel like you have been a setback if you make a mistake on the way. Life is a learning process from birth to death. There will be many mistakes, and also many achievements made in your lifetime. For the love of God, please do not compare yourself to others. I feel as though this is the main reason why we want to be perfect. We see all these celebrities day in and day out with lives that screams "PERFECTION!."

Social Media has led us to believe that the perfect life is possible. That is a huge no-no. Even worse, we beat ourselves up for having the current life we have and always worrying about the life we don't have, instead than thinking about the things we have and cherish every single day. We beat ourselves up over the ridiculous things, whether it be money, or having the ideal body. I feel as though we point out one flaw in our life, and our life is a straight crap show. Everyone has their problems, some way more intense than others. We seem to dwell way more on our cons. We are driven by feelings rather than our thoughts and acts. Here is another saying by Mel Robbins, that I have been taking a lot of consideration to lately. We put our feelings first rather than our actual thoughts. And you know what those feeling usually do? They stop the action because they think into the future. These feelings typically say, "Oh, that is an awful idea; just think of the outcomes!" Feelings will get in the way of everything in our lives, whether it be trying to start up a conversation with a person in a suit and tie, or an attractive woman in the gym. A feeling will also get in the way of improving oneself in the long run. You don't have to be perfect to get the point across or state your opinion on a particular subject. Having courage is much more realizable than being perfect. Just remember your life is your rodeo. Have fun while doing it.

Real is Rare

Every person out there wants to be successful. Every person wants to be beautiful. Every person wants that infamous spotlight that will never stop shining. Every person wants his or her destiny to be flawless. Every person wants to be happy. That is why we have social media, right? We now have that power to make us never feel powerless ever again. When I look at these two words, "real" and "rare," I see many similarities in their context and flow. The way people portray themselves these days has many liabilities. Not just through words, but even in personal photos. Here, let me put it like this, you ever watch that show "*Catfish*"? Well, if you have not, I'm sure you have heard of it. These people represent someone else, an imaginary person, who lures someone into a relationship. Pretty messed up, right? Yea, I bet you feel pretty bad now if you have fished before. So please, if you are part of the current social media craze, at least, portray yourself and not someone else.

Now going into this, I want to discuss photoshop. I know, I know you don't want to hear it and you could care less. Trust me, I feel the same way. But, if you read this far and not glanced at your phone yet, you must be somewhat interested. Right this second, right now, some girl or boy is crying and crying about their looks that are real and rare to find because every one of them is unique. The reason they are crying is that they don't have the looks of someone they envy, or the brains, or the talent. They are searching someone up on the world wide web to look at all of their accomplishments, and contributions to the world. The boys and girls are saying, "Why can't have I his/her life?" And you know what, I don't blame them, not even one bit. I'm sure you can't either. Jealousy has never been such a big part of our lives until right now. That feeling of fame and power is all the human mind wants and craves. And we will go through drastic measures to accomplish that feat.

Remember that word I mentioned in the first sentence of the last paragraph? That dreadful word, "Photoshop." That word "Photoshop?" has nothing to do with editing a photo for someone to look better, hotter, and envious. No, it has to do with what the people make them be. You have to look at the bigger picture, manipulate the word into something broader, faker. No word choice can replace such a vile disease, but let's make one up. How about "Life-Editor" instead of "Photoshop." Yes, I think that is just right. This life-editor mechanism has the power to make anyone one of the most powerful human beings in the world. Whether it be money, real estate, fame, or power over the people, these people are the evilest, even if it doesn't seem like they are. They are advertising themselves to us for us to give into them. And what is sad is that we do give into them, whether it be them selling a product or whatever, they are winning. What scares me the most is the idea of the newest generation thinking that this is the way to live and prosper, through social media and being a human being that is not real.

With all the enhancements put onto pictures, how could one tell if it is real or fake? Technology has bestowed upon us the endless cycle of beauty, day in and day out. It is hard for people to accept the way they are, and now, social media has the power to change that. What matters more, being right or fitting in? Being someone you are not was never a saying back in the day. Even when I was a kid, there was no such thing. People lived their intuitive lives with no followers, just leaders. In today's day and age, people would do anything to go back in time to change a specific event or maybe to be young once again. But, here is the thing, one should never resent getting old, it is a privilege. Yes, I believe that growing old and living your life isn't a right and it never was. Living a life that you carved out is a privilege. And this privilege is your world, and it only exists because only you exist. That world is a self-portrait of you that only has one pair of foot-prints. You set the path on your inner and outer journey. Judgment from others should and will be set aside as you make your way through your world.

I don't want you to get emotional with me now, but what you're doing on that phone of yours that you care so deeply for is controlling your life and who you are as a person. The concept of healthy living is starting to be forgotten. People don't care for themselves in real life anymore, only on a screen. This fake embodiment not only changes the outlook on life in a person but will also deeply impact their mentality. Depression is more eminent, and social anxiety is at an all-time high, especially in the new generation, which is scary to think. I wonder how this will impact us in 50 years from now. Will humans still be the dominant species running the world, or will mindless zombies staring at their phones run the things we used to cherish?

The Broken Science of Judgment

Everyone go to science, whether it be them conducting the experiment or them being the lab rat. How could one define this specific science that goes through all of our minds every day about others around us? We have a judgment for just about every person out there. And I am sure that most of them are negative. Our perceptions of others are so compelled and contorted. We judge people we don't even know, it doesn't matter their gender, age, history, or work they have done from the past to the present. Competition these days is so stiff with everyone around you, whether it be someone who has more muscle or who has a better stock portfolio.

I used to see this girl who would have a negative comment for anyone she saw when I was with her. I remember back in my junior year of college we were in the dining hall around lunch-time. I was concentrating on my food while she was concentrating mostly on her phone, but on occasion, she would glance to see what was going on within her peripheral surroundings. She saw this one girl who was chewing with her mouth wide open, and as I looked up at her, I saw the face of true disgust and repulse. She blurts out, "Oh My God! Close that freaking mouth!" Thank the lord it was not loud enough for the other woman to hear, but loud enough for other tables to look up to see what was going on. Here is where it becomes more intriguing. She sees this one girl walk into the dining hall, and she says, "Wow, Will look at this girl's nose." I can't even lie, it was a bit on the bigger side, but c'mon, that was freaking rude. I would notice this pattern about every day from her. She was a beautiful girl, but the constant pointing out of other's faults was a huge turn-off, really making me second guess my decision with my choice of women. Her continuous judgment of others made me think. Is it because she was jealous? Or just a bitch who hangs out with people to dry? It was to the point that her judgment reflected how serious her issues were.

Do you know how people try to hide their negative judgments about others up inside that brain? Well, she just blurted them out without thinking or filtering. I probably shouldn't be writing this, but her ex-boyfriend physically and most certainly mentally abused her to the point where she was six feet below. She couldn't exactly escape the relationship for some reasons I will never understand, and reasons I will never judge. However, the way she acts around others, thinking she is the hottest and best girl in the room makes you think how lonely this girl must feel. Of course, she won't under any circumstance show her true self; she will always hide those bottled up emotions.

By next year, I will be starting my senior year, and I hadn't talked to this girl since the last day of junior year. Being so close to someone and then all of a sudden just ending the relationship with a snap of finger makes you think, "Damn, now what?" But you know what, life goes on. Anyway, seeing her on campus in September while not saying a word to her since May, was an unforgettable moment. We just walked past each other like we were total strangers. One night at a party, I saw her there, and she saw me. She came up to me and said sorry in a very drunken and distorted voice. Kind of like an apology you would hear out of Larry David, one that has no significant meaning.

As the school year went on, her friends started to see who she was. The true masterpiece of being an awful human being. Having no care in the world except for herself. No wonder she transferred to a different school, what a waste of a year and a half to go to a different school and change your major. Welcome to the beautiful science of judgment.

Life Through A Device

Whatever you send to people they will send it right back. All of those skeletons hiding in the cupboards under your sink can erupt in that tiny brain of yours any day. What will you do then? Run back to your friends? Well, don't you remember the stunt you pulled? Those "friends" are gone forever. It takes discipline and dedication to run life without love. Those who take pleasure in pushing people away and putting people down are right around the corner from that life.

Ah, felt good getting that off my chest. Now, as you see, judging has its consequences. But the science added a new appendage within the past decade; judgment through virtual reality. And this judgment is visual, not just hidden within the complex mind. Everyone wants to get a piece of the action when it comes to the online game. I feel as though this judgment is much stronger, as well. Pictures and even words on social media can rile up every emotion imaginable out there. And being judged makes you feel unwanted. That word "unwanted" is one of the scariest words that ever surfaced in the world. It makes people want to tear off their skin and become a completely different person. Falling apart and going insane soon follows the travesty of that feeling of loneliness. Getting judged over the phone can hurt just as much if not more than face-to-face. Because here is the thing, at least in person, there will be some filtering; over the phone, people don't feel the need to filter their emotions or words towards someone else. The worst part that comes out of this is that everyone on the internet can see what's going on. Everyone is connected, do you think anyone will step in and intervene? I doubt it. I mean c' mon, it's words on social media, how bad could they hurt? But you, the reader are smarter than my mediocre questions. You already know all the dangers that one single comment can bring to those weak-minded. Judgment kills, and it kills every day. It may take away someone's spirit or someone's will to live.

You may be thinking, only bullies pull this kind of crap, but you are wrong. Every one of us, every day, maybe thinking we are just joking around. But judgment is a crime, and it is not believed to be by many. I have a challenge for you, one that you think will be super easy, but when it becomes prime time for a slap in the face with hurtful words, you will repress those thoughts. Go about a whole day without a single negative judgment. Instead, be caring. Buy that newspaper for the older woman, let the person with three grocery items cut you in line, make someone's day rather than diminish it. Be attentive to those in need and not pay attention to their outlying flaws. Pull some chords, and you could complete the day without a single judgment if it bothers you this much then be observant to something else like the clouds, tall buildings or even better than the handy dandy phone of yours! This is a significant step to how the doors of the universe will open for you. Be positive towards the world, and the world will be positive back towards you.

The New Meaning of "Attention"

Ah, that word: attention. Attention has been brought back to life and risen from the ashes with the birth of social media and all of its wicked gloriousness. Remember how we humans loved attention in person? Well, we still do, but now, we could feed that need over social media as well! Yes, I bet you already know what I am thinking. I'm not just looking at women either; it's you too guys. Even if you don't want to admit it, you love that feeling of attention. I mean, who doesn't. Being called handsome or pretty is a great energy booster. Endorphins run through the veins, and it puts a smile on your face — a great and subtle thing to get you through the day. Being applauded for a great act is an extravagant sensation. How about getting 100 likes on a picture you posted on social media? Hmm. Wait, what just happened? Do you feel significantly appreciated? I doubt it. However, in today's day and age, this is the way of people getting attention. Notice how I say, "getting" instead of "receiving." Wow, I just blew my mind by writing that sentence down. Hope I blew yours too. Anyway. People now suck on others to get that feeling of attention instead of receiving. It's kind of like a bum on the street begging for change. The person on social media that posts something is intentionally getting on their knees and begging for attention. The word attention is no more a luxury word, and it has been diminished to a weak word for the pity.

People would do anything these days to get their name heard, whether it be something like consuming loads of alcohol so you could be considered "the guy" at social gatherings or something insanely dangerous like jumping off a roof to body slam a fold-up table. Being intoxicated isn't an excuse anymore for all of these ridiculous acts just for the laughs and awes it will bring to people.

Phones and social media isn't just changing the world around us, but the souls within us as well. We think differently, and we act like someone else is controlling us. We're like puppets in a never-ending show. The spotlight shines down on those with the most attention, while those trying to chase the spotlight may eventually get there, but on the inside, he or she will never reach it. You know why? Because there is no such thing. No such thing as a spotlight on social media or its subsidiaries. Reaching a stage outside of the virtual world is a challenging one, one that will bring you to your knees time and time again. But once you reach that and keep improving, that spotlight will always be shining down upon you. Here is the cool thing about that, no one will care, but you do, and that is all that matters.

What About The Children?

Technology and Education Just Don't Mix

Going off my first meaning of the word attention, I feel like us Americans are getting dumber and dumber as the days go on. Reading, writing, and listening has been downplayed to a more virtual setting within devices. Schooling these days has been increasing with online lectures, where the student can sit in his or her bed and watch the lecture and do the work online. Welcome to the future. Even business meetings can be conducted online. Convenient right? However, I wonder how long our attention spans are through our devices. I feel like whenever I get lost in social media and soon regain myself, I didn't even realize how long I was staring at the screen. The scary part of online lectures is where you watch the whole lecture, and at the end, you say to yourself, "What the hell did I just watch?"

I remember the good old' days when school was simple with just a pencil and paper. All tests and homework were handed rather than submitted online. Yes, the most reliable way to hand something into your teacher was with a handoff. Back in elementary school, I made a fool of myself one day. Well, there were many more times, but I will tell you one example. I was in the 6th grade at the time. The class was about, to begin with, all of my current long term friends and the teacher asks me to grab the projector from the classroom across the hall. I agree and grab the hunk of junk. As I am bringing it into my classroom; however, the wheels get stuck in the pane of the door, and the projector flies forward shattering into hundreds of pieces in front of me. The class, of course, gets a huge laugh out of it while I am stunned and my face is tomato red. I look up to see my teacher with her hand over her face shaking her head, and my mouth dropped still processing the moment. Even though it was an embarrassing moment, I would one hundred percent go back to that time where everything and everyone was much more simple and calm.

Not too long after this, I received my first cell phone at the end of the 7th grade. It was a flip phone, and I thought it was the coolest object in the world.; no need for a house phone anymore and need to write my friends numbers down on a piece of paper. The new world of texting my friends was the latest and most rad thing around those days. Before I knew it, I had the stellar Blackberry in my hands. High school came around, and devices were much more popular where just about everyone and anyone had a phone. Keep in mind the year is around 2010. You may remember the game Brick Breaker on the Blackberry. That game was super addicting. Before we knew it, the Apple iPhone became the mainstream, and everyone out there was getting their hands on one. The new game of social media was upon us. I still believe it is the most competitive game out on the market. It felt like a race to get the most followers, friends, and likes. The game of social media won over society and the people that inhabited it and still do.

With this new fast technology, school systems started adapting to change as well. The age of devices was upon us, and education soon followed suit. Online was a new and integrative way of schooling. Luckily, most of my early education was through handwriting and learning how to write cursive. We now have devices that could write an essay in any format and font we desire. Instead of filling in the blank on a piece of paper, it is now filling in a bubble on a computer, with a simple click of a button. At first, I thought this new way of schooling was very optimistic for a forward-moving world. However, once I got into college, almost everything, even lectures, can be done and seen through a screen. The sense of learning that I used to know and hold was changing, and I didn't like it. Taking online quizzes and even tests on a computer rather than a piece of paper seemed foreign to me. It certainly impacted my learning approaches in many ways. I started to miss the days when I learned information on a desk with students around me. I missed the way a pencil felt in my hands and on a piece of paper. Since lectures were online now, students didn't feel the need to go to class anymore to learn hands-on. I thought that going to class and meeting new people was the whole point of school. Obtaining powerful insight from professors and students was what I built on, and many others apart from my generation. Those before me could probably say the same.

So amid things, I want to know from you guys, do you think education and schooling are better off with all of these changes happening so fast? Will the next generation after us benefit like those before us did? You have to remember this, many millionaires and even billionaires out there left education behind. Instead of finding a job, they created them. That's what separates the middle class from the upper class. Creating and inventing rather than working, is what makes people rich. Taking the risk of following a different path can lead to a full lifetime of success. However, many people will never take this risk because they're comfortable on the safe side of the spectrum. Going out of our comfort zone seems foreign to many of us. Education and finding a job afterward seems to fit right into the agenda if you wish to be successful. Like many of us, education is the same way. They adapt, rather than create. This is what separates the brilliant and the bold, from the educated and anxious. Both can become very successful in their way and carve their path. Speaking of our newest generation, I hope some can get a hand on one of these books. Not just my book, but also any book to expand their current limited capabilities.

Take, for example, Mark Zuckerberg, the creator of Facebook. He is still very young, as well. There is no doubt in my mind that another creative mastermind will come about, probably very soon. However, now that our mind is more enclosed to social-media can something like this be accomplished again? At the moment, where our phones bring out the addictive sides of us, it is hard enough to get out of bed in the morning. I also see in the past that people were a lot more motivated and engaged to get through the day; now, everyone dreads the workdays and cannot wait for them to be over. When a problem arises, we try to avoid it at all costs. Not because we are running away from them, but because our problem-solving skills have decreased with these brand new devices in our hand to solve them for us. The good old fashion way of solving something is now dried up with a new way to solve a tricky problem in a matter of seconds. The way we think and act throughout our day has undoubtedly changed. I don't believe it is a beneficial way to change the way the human mind thinks and reacts. It may be a little too late for my generation, so let us all work on the youth of the nation that is being brought into the tech-savvy world, rather than adapting to it like we still are.

Camp, Where You Find Your True Colors

Maybe you guys remember camping out back in the day as kids. Perhaps you don't want to remember those days. Days of camp will consist of bottled up suppressed memories or memories you will cherish and look back on for the rest of your life. They may have been troubling times, or maybe you found out who your true friends are in life. Whether it is a go-away camp for a month or after school, as a kid, you made other human contacts. You met your "homies" here for life and met your worst enemies here. But what I am trying to get at is that as a kid, you need as much exposure to others as a sunflower needs exposure to the sun's rays. Because you know, a sunflower grows and grows with exposure to sunlight, just as we do from other human life forms. As much as we hate everybody, Monday through Sunday, we need them so we could stay sane. As children, we need friend groups to understand mentally the science of being sociable. I went to a few camps while growing up. I've been to camp back in my elementary school years, high school years, and I'd say even my college years. Going to college in New Hampshire felt like a go away camp where the memories and friends will stay with me forever.

For myself, camping felt like a stressor. I think it was the summer before the start of my 5th-grade year and my parents made me go to a camp for a couple of weeks in upstate New York about a two-hour drive away from Manhattan. I, of course, disagreed at the time, but wouldn't you know it, I was packed and ready to go the following week. It was scary at first because I didn't know anyone. I remember making one trustworthy friend there. His name was Devon. He was in the same boat as me, not knowing anybody and felt out of place. I was always the little guy growing up, but Devon had the same stature like me, so we hit it off quite quickly. I remember the atmosphere of being there. It felt so out of my league because I was a city boy, and now, I am here living out in the sticks for two weeks. Without my mother and father constantly there, I felt out of place and not sure of how to take care of myself. However, all the counselors there were supportive of all our activities. They all participated with the kids to make sure we all had a good time. The point of a go away camp isn't just for fun out in nature but is to grow mentally without the support of your parents. Being independent and making decisions on your own is what develops a child's mind.

I remember being in camp back in the day with my good pal Joey, whom I am still very close with to this day. For two summers, we went to the camp, and at both times, we had a blast. This camp is different from the go away camp. This camp was in the neighborhood over from where we lived. It was good going with somebody I knew, even though when we went outside, the basketball hoops were bent into an oval shape with no lines to tell when the ball is out. It still was a great way to spend a nice day. I remember this one day pretty vividly from camp. We were all outside doing our activities. I'm pretty sure Joey and I were tossing a Frisbee back and forth until out of nowhere we hear a loud female shout. A shout, so loud that you could probably hear the shriek from down the block. One of the campers picked up a rock and hurled it at the girl. The girl's glasses were on the floor shattered and contorted. Her mom was, of course, called but the situation was handled. She was bandaged up, and the camper who had the great idea was locked up in timeout. Now, look at this situation in today's world. You have to remember that camping for me was over a decade ago. Who knows what the consequences the kid that threw the rock had to go through if that was in today's world? The way camps are these days; I would not be surprised if they had the authority on speed dial.

Let's step five years into the future, where I am currently halfway through my high school career; It's the summer in between my sophomore and junior year. My parents signed me up for Rustic Pathways. It was a community service trip to help build a kitchen and playground for elementary school children in Costa Rica. I was thinking to myself, "Costa Rica? A solid two thousand miles away from home for a week, and worst of all no television or internet." Remember, this is around the time when the iPhone is starting to surface into the world. The only other time I had been out of the country was when I went to Italy as a little kid. I was on my way to the airport where my parents dropped me off to meet up with the other wholehearted volunteers ready to stack some bricks and string up a couple of tire swings. However, I soon realized the ones I became friends with on the plane were going to a different site in Costa Rica. So, I'm already thinking, damn just as I was making friends they are already leaving. When we landed, it felt like I was in a completely different world. Instead of exiting the aircraft into an airport, we stepped onto a steel staircase to the airport grounds, and all exited to a bus waiting for us outside the low traffic airport. The drive to the site was about an hour away from the plane and seeing the third world country while listening to my favorite Eminem songs opened my eyes. Seeing how good we all have it back in America compared to a place like Costa Rica makes you think, especially as a snot-nosed high-schooler who isn't grateful for anything. Of course, that's a lie, but you know what I mean.

As we entered the campgrounds, there were three buildings. One was for the boys, one for the girls, and one for laundry and other miscellaneous items. As I set my bags down, most of my roommates for the week were already settled in. Every single one of them was from different places on the globe and had different accents and different ways of living life. I remember this one guy from Virginia, his name was Stephen, and he was a total tool. I didn't enjoy his presence right off the bat. Another was from Seattle and was entering freshman year of high school. He said "hella" in just about every sentence. Another was from London and was in the summer before his college career. He thought he was the bomb.com and was a bit too in love with himself. Most of the guys there I didn't really like, except for Sean. Sean was from Illinois, the state my mother is from. He and I hit it off. He didn't care too much for the other male campers either, which I respected. At this point, it was getting late, and the community service started the next day. I never thought going to bed was so hard until the very first night here. Still very nervous about what the very near future would bring, I had an open mind to meet new people and give a helping hand. I mean hey, I am stuck here for a week. Let's see what these adventures can bring.

The next day, breakfast was served at two long tables where boys, girls, and counselors sat with each other. I couldn't believe my eyes when I saw these girls. Just about every girl there was beautiful, it was like something you see out of a Disney movie. This one girl from Canada was the most attractive thing I have ever set my eyes on. It was like love at first sight. At this point, I was thinking to myself, "I one hundred percent would rather get to know the female volunteers than try to befriend my room-mates." I tried talking to her, and right off the bat, I knew that she wasn't interested in anything I had to say. At this time, I feel as though puberty hasn't hit him me in full circle yet. So eh, that's a bust. However, I got to know this one girl, Meredith, really well. She was from Australia. I couldn't believe that she was from a place so far away from my home in New York. In the other corner of the world, and somehow the stars aligned us to meet in

Costa Rica. Luckily for me, she was with my group for volunteering.

Later that day, we were off to the site where the kitchen was in the process of being built. It wasn't like a kitchen you see in schools around cities, but a kitchen set up like a classroom where students learn how to cook. The kitchen was across the street from the actual school where the kids learned to read and write. The kitchen classroom was on farmland, where spinach, tomatoes, and a variety of vegetables grew. Further away from the unfinished kitchen was a barn with lamb and pigs. I couldn't believe how good the vegetables tasted so fresh and pure. Our job for the kitchen was mainly the floor, filling the empty spaces with ceramic tiles and laying some cement foundations. We also planted seeds and watched the plants grow throughout the week.

I must say though the best part of the trip was playing with the kids and getting to know their culture. We played soccer just about every day. I am awful at soccer, so I just passed the ball to the swift Costa Ricans. It was amazing how much energy they had; it was like they could run for days on end. Whenever I played goalie, I would get violated, and they would score, at least, nine out of ten times. Athleticism was strong with the young ones in this country. Unlike our country, where the young ones look unhealthier than their elders, it seems.

There are lots of fun memories from that adventure, but one that will go down in history was one of the last days in Costa Rica where the group went zip lining. Yea, yea you may be thinking, "Oh wow, zip-lining how fun!" In a very sarcastic tone. But this specific zipline started at the top of the Arenal Volcano. From the very top, you could see just about the whole country on a clear day. Once again, I was shaking in my boots because of the immense height. It was like the entire group was used to this stuff because they all seemed confident and ready to take on the enormous volcano. But what an experience it saw, the countryside and jungle below my feet while zooming by on a skinny line. I would like to see a camp do an activity like that in today's world. I would say that maybe 10% of the kids would sign up for that, rather than the 90% that would back a decade ago.

It feels weird thinking about this because you don't see those kinds of camps anymore. Well, camps are the same, but the kids aren't. You already know what the past was like; being outside was the greatest feeling in the world as a kid. Whether it was playing a game or any sport, now, children would rather stay inside. Working at a camp as a counselor opens up your eyes to see how much kids have changed from a decade ago. They could play on that device hours on end, without even noticing a minute passed by. I would be a hypocrite, however, if I said we adults didn't do the same. I feel like the outdoors became more foreign as the technology age struck us like a lightning bolt. So, the real question I will ask you is, can youth revert back to the ways where a device didn't control life?

Children Would Rather Hang Out With Siri

Our youngest generation has been born into a world already filled with technology, where parents and older brothers and sisters already have a mobile device. I see children these days as young as about seven years old who are already using a phone daily. These children are already developing the nasty habits that all phone users have that eat up all the time in a day, and not to mention, our well-needed sleep. However, there is still time to save the young one's from falling suit of the ones just before them. Parents need to act, or their children will fall victim to the social media game that will continuously trap them.

A while back, I called up my younger cousin who lives somewhere in Chicago. I asked her how school was going, and all she gave me were the same few word answers every time. "Going great," or, "Doing good," was just about the whole conversation. So now, I am on the verge of hanging up, but I wanted to make sure we had somewhat of a conversation. So I ask, "You do anything with your friends lately?" She said, "We saw a movie last week." To myself, I was thinking, "Wow, I got a full sentence out of my younger cousin. I reply, "Which movie?" She responds, "Some romantic movie." I ask, "Was it at least good?" She responds, "I can't remember, I was on my phone the whole time with the sound off and the brightness down." I was shocked even to hear that. I mean I am not surprised, but c'mon in a movie theatre while the flick is playing? So, I ask "Why even waste the money to go see a movie if you are going to be on your phone the whole time? I bet your friends were pissed at you." She says, "They were all on their phones too." I was thinking to myself, "So, this is what the younger generation does when they are with each other?" Hanging out to them may seem played out. Hanging out on social media is part of the new wave. It's to the point where the new generation would rather hang out with a phone and make people believe than actual friends.

Children games have certainly changed over the years. I remember back in elementary school playing cards such as Pokémon was a huge thing. I am sure most of you have already heard of the app Pokémon Go. It came out in the summer of 2016, and just about everyone and their grandmother were out and about trying to catch virtual Pokémon.

Hey, I can't even lie, this is a great idea. Not only did it grow in popularity, but physical activity was skyrocketing at this time. Since the only way to catch these Pokémon on a screen was to get out of the house and explore, physical activity was inevitable. I never downloaded it. But now that I am thinking about it, I believe this is one of the very few applications phones I have to bring that may be beneficial to our health. However, one has to look before blindly walking into a street or a pond for that matter.

Phones have inevitably come a long way over the past couple of decades. The new generation grew up with the web expanding to all corners of the nation, but for the newest generation, the internet and social media is part of their lives every second of the day, and in hand when it comes to bedtime. Phones have radically changed children's sense of overall health in both the mental and physical aspects. Here is a scary analogy; you know how cocaine is the rich man's drug? Well, owning a phone, it has no curtail to whom it can belong too. Richest of the rich or one the poorest souls to walk the earth a phone. And as I said earlier in this section, our attention spans are decreasing with every minute we spend on a phone. Social interactions are deteriorating among the newest generation, and social media interactions are rapidly increasing.

I mean hey, at least, the kids now will be safe physically right? The chances of them walking into the woods or get taken away in a van are slim to none since all they do now is lie in bed with their two hands smothering their phone. However, staring at this mystical device for hours on end with no human contact can cause some severe distress, such as eye redness, or a headache in the short term. Long term is a different story, and to tell you the truth about those children on their phones right now, if they were to read this next section, they would completely disagree because the exuberant feeling of being independent has decreased since the past generations.

Can you believe that children and young teenagers these days can become depressed? I mean, of course, you have heard the news of all that crap. The answer doesn't even lie in bullying anymore. A chunk of it does, but a new kind of "bully" has entered the school. And this bully isn't only in your daughter or son's school. They are in everybodys. It is a psychological bully that isn't even there it is invisible and untouchable. Here is the solution to stop that big ugly bully. The child can't do away with it on their own. The parents need to intervene and make rules about the time spent on a phone. I'm sure every child out there has had this happen to them before where the parents and schools have set up rules to stop the usage of phones and bring them down to a minimum.

Here is what will get the children's heads spinning. Physical health may or may not be affected, but the psychological consequences outweigh the physical ones by a ton. The newest generation may go down in history as the most depressive and anxious group of people to ever walk the Earth. It is hard to believe that the latest technology in this world has the potential to make a child depressed. We all take this new world for granted where we don't even acknowledge the ground we all walk on or the blue sky filled with puffy clouds. The way children are heading as of right now; we are looking towards a massive downward slope. Life isn't meant to be spent unhappy and locked up behind the cell of a phone with no windows, just a black mirror.

Not too long ago children couldn't wait to get out of the house, whether it be for a weekend or completely moving out from the roof which their parents provided them. Independency and wanting to make a positive change in their early lives were at an all-time high. They were taking responsibility for their actions and were being mature enough to find a job to help pay for school and other expenses their parents have been giving to them for so long. Teens today don't feel the need to leave their room anymore. But is it just because of phone and all of its apparatuses? This seems a bit far-fetched for me. It is as though they rather stay home because it feels to them like a haven, with no danger or risk. I mean hey, what could be so wrong about this? The children get to spend more time with their parents, right? Well, you remember how I had that conversation over the phone with my younger cousin, consisting of just two words answers? It goes the same for conversations with parents.

However, that stay at home feeling gets old quick. And once it gets old, children, myself and probably you guys can get lonely. The sad part is, even when we do go out and try to have a good memorable time, phones surround us with the constant documentation of you and your friends having a supposedly good time. I say "supposedly" because I feel as though we don't even go out to have a good time anymore. Don't get me wrong, we certainly do, but our main mission is to record the event. Not for the individual, but for others to see. This documentation gets put on social media, where most of us stuck at home can look at the pictures, and this will onset that feeling of loneliness even more. Not even just in children and teenagers, but in all of us.

It is no wonder why kids these days are becoming more and more brats and unappreciative for all the things they have and receive daily. Phones deprive them of just about everything you can think of, ranging from sleep to having a good time with friends. To them, life revolves around their smartphone, not even thinking about themselves when it comes to health both physically and mentally. However, as more time goes on the more, we all learn about the potential dangers, consistent phone usage has on us, on our kids, and the kids after them. So, let's decide. Are we ready to make a move to stop our constant social media streaming? You have to remember your actions on the phone have a direct correlation with everyone around you. The phone world has developed around us and swallowed us whole. But the youngins' still have a chance not to follow suit. With their nimble young minds, they could be the generation to not worry about how many likes they get on a ludicrous post. They could be the generation to remember what they had for breakfast and be appreciative of it. Ah, yes, the youth of our world. You may hate their guts, but you have to love them too.

Youth Is the Answer

You are probably saying, "Youth?! Ha, give me break; all kids are selfish and privileged these days." Hey, I can't even argue with that. I was a camp counselor for a few summers. It was always a good time, but you know, there are always those few kids out of the group causing a ruckus and constantly being a nuisance. I always found it odd though that one of the first rules in camp, and I am sure in many others, devices were not to be touched until a certain time. Similar to a break at work. Of course, these kids would be sneaking in some playtime on their little handheld devices whenever they weren't being watched. It was like sneaking in a cigarette behind closed doors. It's as if these kids were addicted to all of the gizmos within a phone. Most of the kids even had a social media account. I can't even lie, about half of them had more "friends" than me, and I had a solid ten years on these kids. These kids would always compare to see who had the most "friends," like a game of popularity. Popularity isn't only the jocks or cheer-leaders anymore; it could be anyone or anything.

Recently, my co-worker Angelina, came up to me saying how loud the children were behind the nursing home, the place where I'm employed as a dietary aid. I replied, "Every time they get out of school, it's like they are rejuvenated with energy. They are always playing some game with each other." She then said, "Yes! They were playing grocery store," I replied, "Grocery store?" She replied, "C' mon, you know, with a customer with fake groceries and a cashier and such." I shrugged, but in my head, I started to think that these kids are still playing with toys and sticks rather than sitting on their asses staring blindly into a bright screen. She soon after said, "You don't see that too often anymore with children." I replied, "And why is that?" Already knowing the answer. She says, "Because of phones and all of those other techy savvy devices." I could not agree more.

I told Angelina that I mentioned her in this book, and she thought it was a great idea. She dislikes the social media craze so much that she set up rules in her house about "phone time." It reminded of a camp, and she was the head counselor. One rule in particular that I believe should be the cornerstone in every family house is about using phones at the dinner table. She makes sure that no one has their phone on them or insight while engaging in a family dinner. I couldn't agree more with her. I was even surprised to hear that they still had "family dinners" in general. I put those words in quotes because, in today's world, families don't try to have these anymore. The only times where this may happen is during certain holidays, such as Thanksgiving or Christmas.

She went on to tell me a story about her son. Her son is around my age, and she says that the kid is always on his phone. Just like many others around my age, boy or girl, we are always on that thing. However, he plays guitar. One day, out of the blue, he told Angelina, "You know what, I am going to try to write a song." Angelina in disbelief says, "Hey, there's a good idea." And wouldn't you know it, he set up his station in the living room with his guitar, and an empty note stand trying to find the right rhythm. She was shocked actually to see her son in action.

Another day, she saw him nostalgically looking at baby photos of himself and looking through the innocent photos by flipping, rather than swiping on a device. The photos that are always kept in a place of worship rather than within a phone, where pictures are usually forgotten and have no meaning whatsoever - photos that consist of random selfies, or a picture of a handsome omelet. Tangible photos will always be cherished because they are kept always in a nearby shoebox instead of the depths of a sim card. It made me think that the babies being born right now are kept in albums on the phone rarely ever looked back at. One can compare these unique stories to the children playing their games behind the nursing home.

I started to watch these kids behind my job more and more, just observing the activities they would participate in. I could also tell that they didn't have much due to their outdated toys, but they still made the best out of it. It looked like they were having a lot more fun than the children who are constantly looking at a device. It brings me back to when I was a kid in elementary school. Just about every day after school, and especially on the weekends, I would play with the other kids on my block. We would play a variety of sports, ranging from hide 'n' seek to kickball. But you know, as we grew older, we stopped playing outside. Not just because of our growing age, but the growing age around us.

I also noticed that my block started to become more congested with cars. Back in the day, cars filled the driveways, but now, they started to fill the streets. The population in my neighborhood nearly doubled in the past decade. Back in elementary school, there was a vast concrete parking lot where we would be allowed to play during recess, where just about every day I and the rest of my grade would play kickball. As I drive pass that used to be concrete field today, that seems to be a distant memory because children now are secluded to a smaller play area for recess. Not only did cars become more abundant, but houses as well. The school gave up about half of the parking lot to real estate where several houses are now being built up. I could see why kids didn't want to play outside anymore. Congestion fills the streets - and not to mention the smell. The smell of pollution was becoming a massive turnoff to being outside. However, this is more towards the present day. Playing inside became more popular about a decade ago where just about everyone was adapting too.

Playing outside is not normal anymore. The fun and games are over. It is all about the inside games and let me tell you; I became addicted. Just before and into high school, my life revolved around videogames. After school, I would play for hours on end. Pudginess started setting in, and I couldn't have cared less about my appearance. Do I regret it? Yes, but it was certainly a learning experience to achieve where I am now. I have successfully graduated from college with a bachelor's degree in Nutrition Science, and haven't touched a videogame controller in months. It feels good not to stare at a screen for hours on end. It relaxes not only your eyes but your mind and brainpower as well. Who else wants more stress in their lives? I'm as sure as hell , I don't, and I am sure you don't either. That's why when I observe these kids behind the nursing home, I notice that they are in their world and don't have a worry to stress over. It makes me think, why can't we adults feel that joy like we once did? It is rare being constantly happy, and if you are that person, you somehow beat life. Life is hard to beat, and when you try to beat it, it will beat you down even further. However, this is a good thing, because you will become stronger as a result.

I believe the answer lies within the blindness of the parents. As you readers already know, the reason why the youngest generation has a phone is because of the parents, which makes sense, because in today's day it is needed for constant contact. Many parents today are also addicted to their phones, always on social media, and missing out on the growth of their child, both physically and mentally. Luckily for me, social media and the obsession of always being on the phone did not exactly happen until I was halfway through high school. I noticed lately that my mother is always on her phone. She comes home from a long day at the office, and all she wants to do is relax. Which I could respect, but even when events are happening, such as the simple act of going out to dinner, she is still on that thing. My dad takes notice in it too, always bickering that she is always on that thing, eating her life away. Well enough talking about my parents, they already did their job of raising a kid. They should give themselves a pat on the back because I somehow turned out alright.

Do you know how early 20-year-olds are always on their phones today? Well, now think ten years into the future. They might have a kid; will this distraction of the phone be a constant bother? Parents have the power or should have the willpower to stop streaming social media easily on their own. There is one simple reason for this; in the past, it was different. They didn't have the constant luxury of opening a phone, but now, their children coming into a tech-savvy world are drawn to picking up a phone because everyone else is doing it. I feel like "Social Cues" have certainly adapted to something stronger and more influential into today's day and age.

The Impeccable Challenge!

Getting away from all my wild and outlandish theories, I have a challenge for you, a challenge that will test your willpower. I believe that looking at your phone, checking for any texts or checking social media is like a drug. No, not the drug that just popped up into your head, but more of a physiological trigger that makes you think checking is a reasonable thing to do. It's a habit, a habit that is hard to break, such as biting your fingernails or smelling your fingers after you picked your ass. This challenge can last as long as you want, or as long as you can. Do it for, at least, a single day, weekend, or a whole week. Go without checking social media. Here is what I recommend, no need to delete the apps, but move the apps to a separate page on your phone, so the temptation to look is decreased. I asked others in the past, including roommates and even parents, but they failed. I understand it's a habit, even if you check by mistake, realize it, and correct it. Turn your phone off if you must, or maybe that's pushing the boundary now into your personal life. Give it a try, I believe in you. And think about it this way. Once Monday comes, you will have a lot of catching up to do on social media! You will be busy for hours, staring mindlessly into your phone.

Now listen here, being on that phone, even if it's for a couple of minutes is holding you back. This is why I want you to attempt this challenge with the eagerness to succeed. It is not a simple task. Even though it sounds like one, it is easy to slip. Slip back into that brain dead mindset of looking blankly at a screen. Think of the screen like this. This screen, encased in your phone, is mocking you. Mocking you every time you look at it and laughing in your face like a bully, a big ugly bully. Do you want this bully to win? Every time you pick up that phone to look up pointless crap, it wins, laughs, and spits in your face.

The second part of this challenge is to observe others, no, not from behind trees or in a van, but nonchalantly 'people-watching.' I'm sure you have done this already once, twice, or an uncountable number amount of times, but look for the specifics, and by that, I mean how often or how long they look at their phone. Take note on how long they are on the phone and how many times they take a glance at it. You will be surprised at how many times they will peek at their phone for no apparent reason and put it back.

The last and most important part of this challenge is to observe your surroundings and take in the moments that are meaningful and the moments you usually forget at the day's end. This is the part where synchronicity hits you like a slap in the face from the universe. Take it all in, acknowledge, remember, and record. Most importantly, believe.

Here is a little bonus challenge that can be helpful in accomplishing this feat. When you are just about ready to go to bed, put your phone in the other room. Don't give me the bullcrap of, "But I need it to wake up in the morning." No, you don't, make it loud enough to hear, or even better buy one of those old fashion alarms. The reason for this is because I believe the most tempting time to pick up a phone, or even anything around you is when you are sitting or lying still. The light that comes from a phone will also disturb your sleeping cycle. Mel Robbins goes into great detail about this particular habit before bed and just waking up from a deep sleep. I recommend checking her book out, called "The 5 Second Rule." If you guys could pull this off, the rest of the challenge should be a cakewalk.

Make this a habit, rather than a challenge. Here's the easy part. Now that you accomplished something that many cannot, you are ready to make this into a lifestyle. Whenever you meet someone new, ask for the person's number, instead of his or her social media account. Trust me; it feels much more rewarding getting those digits, rather than the person's fake personality on a screen. When out at a social gathering, restrain yourself from looking at your phone. I notice whenever I go out, I always see people on their phones instead of having a good time. Whenever you are not attached to your phone, you will stand out of the crowd and look much more approachable. Life will change dramatically when you take this step. The experience you feel daily is much more valuable than those meaningless experiences you have within your phone.

I feel like these days; it is getting harder and harder to stay happy for a prolonged period. What I am saying is that your happiness can be taken away at any given moment, and sometimes, it is hard to get that back. I wish I knew the answer to forever happiness; maybe some people do. But I can tell you this, and I will tell you over and over again until you get this message imprinted in your mind. To unlock it, you must look around and take note. It is as simple as that. Every day is precious to better yourself, don't waste your life away on a phone that has meaningless value.

In Conclusion

Wow, you did it. Pat yourself on the back. You are now on the final paragraphs and ready to spread the word, not my word, but your word, a word that describes you and all your values that are ever-growing and evolving.

One could not go through life and say they experienced it if they spent half of it looking down into a screen. All of your imaginary friends on that darn screen aren't allowing you to live your life the way you want to live and thrive in it. Live your life to the fullest and keep that distraction off and in your pocket. The world is your oyster, not your phone. Meet people instead of sending out friend requests to those you said one word to in real life. Make your moments instead of sharing them with others. Why would you care what others think about what you have done or accomplished? Take it in and embrace it, don't share it. Because by sharing it, that moment only becomes less valuable. The moments you feel and receive every day are priceless.

Incorporate and emerge yourself into the ever-mysterious world. Let it grab you and lead you into the unknown. Find the hidden traits within you that are locked up behind brittle bars. Embrace your surroundings and always show love and character to those around you in your daily life. Always remember not to give too much thought to the complicated things in life because there is always a more simple solution. Find those simple things in life and expose them.

I want to share one last thing with you guys before you head out on your journey. I found an excerpt out of the New York Daily Post, and I have to share the wise words it says. Firstly, it is an article by Eric Fettmann interviewing Heather Wilhelm about how there are real addiction problems regarding smartphones. It mentions how humans are starting to lack basic skills such as concentrating, conversing, and of course, being happy. But the quote I want you guys to ponder about is an actual quote by Confucius. Heather says, "He who conquers the smartphone is the mightiest warrior," then adds, "OK, fine, he said, "He who conquers himself." She then remarks, "Isn't that pretty much the same thing?"

Acknowledgments

How can I not talk about my main man Jeff in the very first line of this section? Shout out to him for teaching me his ways; a huge inspiration throughout my last year of college and many more years to come. This book couldn't have even been started without your wisdom. Thank you so much for your guidance. I greatly appreciate it and always will.

Secondly, I have to acknowledge my friends and family. Maybe a cop-out answer, but if it wasn't for them constantly on their phones all the time, the idea of writing this book would have never come to my mind.

Making the stride to go away to school. Never would have thought as a kid to be going to college in a different state completely away from the city, my family, and friends. I hugely appreciate the consistent encouragement for me to make the bold move.

The people I worked with throughout the years. They are always keeping me on my toes. Giving me wise advice and the experiences they went through, throughout their lives.

Ah yes, my phone, my parent's phones, my friend's phones and, of course, your phone have led me to write this book and share it with you mindful readers. I can't even lie I wish I counted how many times I picked up my phone while writing because it would have been an uncountable amount.

One last shout out goes to the fabulous editor, Alyssa Falconi. Thank you so much for putting time and effort into the book for fixing my silly mistakes.

About The Author

William Civitillo was born in 1995, right before the technology boom. He grew up in Queens, New York. His interest grew about life and health when he was diagnosed with Celiac Disease at the age of 13. This not only changed his life dramatically but also led him to study to become a dietitian and motivational coach. By the time of his senior year at Keene State College, he met one of the most influential and outgoing people out there. Coach Sarri, one of Will's college professors, taught him the lesson of synchronicity and how life is about the simple things and to always be grateful with a smile on your face. This simple but outlandish lesson not only taught Will to be more involved with life but also led him to write this book. Now, Will was never that much interested in literature, but this particular subject led him to go beyond his horizons. Horizons that he didn't think could be reachable.

William wanted this book to be something more than just putting the phone down; that is only a small piece of the puzzle. He wants everyone to take life to the fullest and embrace every moment with full clarity, even if it is a negative one. Now that William is a dietitian, he not only teaches people how your food choices can affect your quality of life but also how one carries him/her self. William's extensive knowledge in both the human mind and digestive tract can lead someone to bliss and forever happiness.

Life Through A Device

www.ingramcontent.com/pod-product-compliance
Lightning Source LLC
Chambersburg PA
CBHW060603200326
41521CB00007B/654